Scientific and Technological Communication

Scientific and Technological Communication

BY

SIDNEY PASSMAN

UNITED STATES ARMS CONTROL AND DISARMAMENT AGENCY

PERGAMON PRESS

OXFORD · LONDON · EDINBURGH · NEW YORK
TORONTO · SYDNEY · PARIS · BRAUNSCHWEIG

Pergamon Press Ltd., Headington Hill Hall, Oxford
4 & 5 Fitzroy Square, London W.1
Pergamon Press (Scotland) Ltd., 2 & 3 Teviot Place, Edinburgh 1
Pergamon Press Inc., Maxwell House, Fairview Park, Elmsford, New York 10523
Pergamon of Canada Ltd., 207 Queen's Quay West, Toronto 1
Pergamon Press (Aust.) Pty. Ltd., 19a Boundary Street,
Rushcutters Bay, N.S.W. 2011, Australia
Pergamon Press S.A.R.L., 24 rue des Écoles, Paris 5e
Vieweg & Sohn GmbH, Burgplatz 1, Braunschweig

First edition 1969

Library of Congress Catalog Card No. 74-91466

Printed in Great Britain by A. Wheaton & Co., Exeter

The judgments expressed in this book are those of the author and do not necessarily reflect the views of the United States Arms Control and Disarmament Agency or any other Agency of the United States Government.

08 006631 3

TO MY WIFE MARGOT,

AND DAUGHTERS JANE RHONDA AND NANCY LYNN

Contents

Preface

IT HAS become quite fashionable to talk about the "information explosion" and the related problems of scientific communication. Unfortunately, there has been far too much talk and too little analysis, experiment and ameliorative action in this field. It is hoped that this work can accomplish more than yet another contribution to the burgeoning dilettantish literature on this subject.

My objective in this work has been to try to get at the fundamental aspects of the elements and media of scientific and technological communication and to describe the critical issues involving them as well as the opportunities and techniques for exploiting them which hopefully could aid both the "users" and the "handlers" of these important resources.

It has been my privilege to be associated over the last dozen years with various high-level scientific and governmental committees that have attempted to review the problems in this field and to make recommendations for coordination and improvement. This included serving as a member of the "Crawford Task Force" for the President's Science Advisor which was asked to recommend a government-wide plan for action, consultant to the Weinberg Panel of the President's Science Advisory Committee, concerned with the Private Sector, and as a member of the "Licklider committee" to review the progress of the previous groups and government action up to 1965. In addition I have served as an Editor of an International Scientific Journal (*Infrared Physics*), Editor of a series of scientific books, and on various scientific groups concerned with holding symposia and publishing proceedings. Most recently, I chaired a panel for the Federal Council for Science and Technology's Committee on Scientific and Technical Information (COSATI) on "The Role of the

Technical Report in Scientific and Technological Communication", which investigated the interfaces and issues concerning the media of journals, books and reports.

All of these activities provided me with a great visibility over the various problems and activities in the field of scientific communication which I felt a fiduciary responsibility to record and make more generally available than the immediate recipients of the subject reports. This book is the result of that effort. In preparing it, I have drawn heavily on the material accumulated in these efforts and am indebted to all of the people who have been also involved in them and related activities of the Federal family of committees. In particular I would like to express my appreciation to my past colleagues on the various committees mentioned above.

In addition, I have tried to utilize the insights and relevant findings of the various authors who have contributed to the collective wisdom on this subject. My thanks go out to all of those who have kindly granted permission to quote the pertinent materials as indicated in the text.

I must hasten to state, however, that I alone must bear responsibility for the various opinions and prejudices which are liberally distributed throughout the work, and for the particular emphasis and structuring of the material.

Washington, D.C. SIDNEY PASSMAN

CHAPTER 1

Introduction

OUR "post-industrial society" is characterized by a great impatience to make progress in our time in all of the dimensions that matter to us. In support of these societal goals we have come to expect great things from the vast enterprise of science into which we now invest a sizeable fraction of our resources of funds and outstanding manpower. If these efforts are indeed to realize their full potential, it will be because of synergistic effects provided by the coherence of the works of many individuals from different technical specialities and from many laboratories and geographical and even temporal contexts. The matrix which binds these efforts together and which enables the entire effort to flourish, is called *scientific and technical communication* and involves a host of media and intersecting networks. Some of the elements of this system can be planned and improved by means of the opportunities offered by modern communication facilities. But other aspects involve subtleties of intellectual processes and innovative thinking which continue to defy orderly analysis and make scientific creativity an exceedingly inexact and poorly understood discipline.

This work is an attempt to analyze the scientific communication process and to describe and critique its elements so as to provide two things:

- a guide to these elements for the benefit of scientific and technical practitioners, and
- a critique and framework for evolving an improved system for the private and government sectors of users, generators and managers of the communication system.

In recent times we and other nations have adopted a very pragmatic view towards our scientific and technical establishments. Thus science and its twin, technology, have become a big and important business, intimately involved with the most vital and significant aspects of our lives, closely related to our security vis-à-vis our adversaries, to the satisfaction of our needs and wants, and to the understanding and control of our environment. The stakes in the success of the business of science and technology are high, for individuals as regards their success and approbation, for institutions as regards their financial and organizational survival and well-being, for states and geographical areas as regards their development and prosperity, for nations as regards their security and economic viability, and indeed for mankind as a whole as regards its enjoyment of a good and healthful life on this planet and in exploring the rest of the universe.

While not questioning the high motives of individuals and organizations to continue to pursue the will-of-the-wisp of the understanding of nature and its orderly but enigmatic workings, it is necessary to keep in mind that these "business aspects" of science are exceedingly important both as a forcing function in supplying the resources for and the directivity of the technical enterprise and in the role of boundary conditions constraining the types of efforts and the communication processes. Whether this is intellectually satisfying or not, the facts of life of modern science and technology are that scientific results have become *proprietary* to the purposes of individuals, institutions and governments. As President Johnson stated on the occasion of signing the State Technical Services Act in 1966, "the test of our generation will not be the accumulation of knowledge . . . our test will be how we apply that knowledge for the betterment of mankind".

These existential considerations must be borne in mind as we discuss the evolution of the technical communication process, since, as we know, it has become more and more of a crucial element in the business practice of science and technology and indeed has become a considerable business in its own right (Machlup, 1962).

In an almost Parkinsonian manner, Science and Technology seem to expand and grow consistent with the resources available for their support. Although there are currently some concerns over priorities of effort, the total scientific involvement of the nation continues its exponential growth. D. J. de Solla Price (1967) refers to this growth as the First Law of Research on Research:

> The size of Science as a function of time exhibits a regular exponential rate of growth, holding for periods as long as 200 years with a doubling every 10 to 15 years. One gets about the same rate whether you count men or scientific journals, or the papers published in them. Rates vary only a little from field to field of science, from country to country.

Greater and wider participation in research in all phases of public and private endeavor, have led to the phenomena characterized by the term *information explosion* or *flood.**

Science is not unique in this regard. Thus, it is observed that "in governments, business enterprises, political parties, labor unions, the professions, educational institutions, and voluntary associations, and in every other sphere of modern life, the chronic condition is a *surfeit of information*, useless, poorly integrated, or lost somewhere in the system" (Wilensky, 1967).

This growth and increased scale of our scientific and technical effort places greater and greater demands on all of those involved in the process, from the problem of the novice to assimilate enough of the growing corpus of knowledge to bring him to a productive position as a practitioner and contributor before hardening of the intellectual arteries sets in, on up to the problems of governmental sponsors of technical programs to assure that their programs properly build upon previous accomplishments and make use of existing knowledge to accomplish their important objectives, in a cost-effective and timely manner.

The great and growing mass of scientific and technical information produces pressures and frustrations for all portions of the communication system. As pointed out by Kent (1966), this problem has a number of degrees of frustration:

> 1. The physical impossibility that an individual scientist or scholar in any field can read and remember all the literature that has a reasonable probability of being useful or interesting at an unspecified future time.

---

*It is interesting to observe the changing metaphors that have been applied to the rapid growth of knowledge. What was originally commonly referred to as an information "*explosion*" has now been characterized as exhibiting more of the qualities of a *flood* (after Licklider, 1966). This follows Garvey's description of the "flow" of information: "The river of information gushes along, carrying a volume of scientific knowledge so great that the scientist is swamped" (Garvey, 1967).

This analogy also reinforces Dr. Weinberg's metaphor comparing the information handler to the role of the ancient Persian irrigation-channel managers (PSAC, 1963).

2. The economic impossibility that an individual or his organization can process for later exploitation, a major part of the literature that exhibits probable pertinent interest.

3. The mechanical impossibility that traditional library tools can cope effectively with the detailed requirements of research workers for information of precisely specified relevance.

More basically, an effective scientific communication system is essential to the unity and integrity of science itself. Thus, after Weinberg (PSAC, 1963), it is important to recognize that:

> science and technology can flourish only if each scientist interacts with his colleagues and his predecessors, and only if every branch of science interacts with other branches of science; in this sense science must remain unified if it is to remain effective. The ideas and data that are the substance of science and technology are embodied in the literature; only if the literature remains a unity can science itself be unified and viable.

One of the significant features associated with the growth of science and technology is that characterized by the increasing complexity of interrelationships among its elements. It is one of the attributes of the scientific method that the individual workers operate by a "divide and conquer" philosophy, known as specialization. It is through this fragmentation process that their strength and pressure can be brought to bear upon a small enough facet of the universe to yield to their probing and understanding. On the other hand, the demands of the community as a whole, whose goal is to learn as much as possible, as quickly as possible about the universe, so as to exploit it for the manifold objectives that society seeks from science and technology, must be directed towards a cross-coupling of specialties and a complementarity of findings. It is this duality that often results in what appears to be a conflict of attitudes about the information crisis. For the frontier specialist, who copes with the growth of information by closing down his "entrance slits" to information by means of an increasingly narrower specialization, he may see no great problem (by having closed his eyes to it). But to the individuals who wish to make use of information across disciplines (viz. engineers) or to those responsible for a project oriented view of science in terms of meeting national objectives in an efficient manner, such an escapist attitude cannot be condoned.

In a very real sense, vis-à-vis our adversaries, our environment and our own amoral technical progress, we are in a race for survival. This concern by and on behalf of society has been well stated by barrister Oscar M. Ruebhausen in his Foreword to Alan F. Westin's *Privacy and Freedom* (1967):

> With each passing year, however, the pace of obsolescence in old knowledge seems to quicken. New data, new concepts and techniques seem to press upon the structure of our lives with ever accelerating force.
>
> Survival, it is clear, depends on the rapidity with which such new knowledge is mastered. This is an axiom for industry and commerce, as well as for biologists. It is no less axiomatic—only somewhat less explicitly accepted—for our social and political institutions.

It is for these reasons that there exists so much concern at various community levels of aggregate responsibility, with the information problem. Thus professional societies, congressional and executive committees, and international bodies are focusing on the existing problems and the opportunities for the future of developing and employing improved information handling techniques.

But the problems of information are complex and subtle. Ofttimes it is better *not to* have information if it is misleading or in error. Will Rogers' classic humor on this theme is just as applicable to scientific areas: "It isn't only what you don't know that hurts you—it's what you *know* but that *isn't so*."

There are also the phenomena referred to by the term, the "social functions of ignorance" and the unanticipated consequences of purposive social action, of which Merton (1961) has written. These remind one of the old adage: "he didn't know that it couldn't be done—so he did it." But on balance we would have to say that information is a "good thing". Even the knowledge that the information system is purposely ignoring certain inputs as being unreliable is important to progress.

Because of the importance of the scientific literature to the integrity, correctness and viability of the scientific and engineering process, the scientific community must be on guard against encroachments from whatever direction which interfere with the proper functioning of the process. There are various interest groups—lawyers, legislators, businessmen, publishers, government "bureaucrats", etc.—constantly trying to exert pressures of one sort or another on the communication process. It behooves the technical community to observe the watchword—"Eternal Vigilance is the price of liberty". Intrusions and erosions against good communication practices must be resisted wherever possible. In the final analysis, the initiatives of scientific inquiry and the methods of communication and the establishment of quality and value judgments, must be left to the choices of the scientific community who alone are qualified to act in this domain.

CHAPTER 2

The Research and Engineering Process and Technical Information

THE ennobling image of classical science is perhaps best illustrated in the traditional concept of an investigator freely sharing his observations with his academic peers and seeking out their detailed review and criticism so that his work can profit from the self-correcting feedback process which is often held up as the hallmark of the "scientific method". The scientific literature itself has represented the evolving fabric of the collective scientific intelligence, transcending individual minds and national and temporal boundaries. As so well described and elaborated on by Professor Hagstrom (1965) in his book, *The Scientific Community*, the scientific "paper" is visualized as a "contribution" tendered to the scientific community, the sole recompense for which is the recognition of that community. As has been observed by Merton (1963):

> Rather than being mutually exclusive, joy in discovery and eagerness for recognition by scientific peers are stamped out of the same psychological coin. They both express a basic commitment to the value of advancing knowledge . . . it need not be that scientists seek only to win the applause of their peers but, rather, that they are comforted and gratified by it, when it does ring out.

Overhage (1967) has nicely stated the classical literature tradition:

> The public printed record of the results of scholarly research is the universal device that transcends the barriers of space and time between scholars. It makes the most recent advances of human knowledge accessible to students and scholars throughout the world. Wherever there is a library, any person who has learned the language may participate in the outstanding intellectual adventures of

> his time. The same record extends into the past; through an unbroken sequence of communications, the scholar of today can trace the origin of a new concept in different periods and in different countries. By standing on the shoulders of a giant, he may see farther.
>
> The wide availability of the record is one of the guarantees of its soundness. In science especially, truth is held to reside in findings that can be experimentally verified anywhere, at any time.

In view of the wide involvement of science and technology with all phases of our society, it is important to note that there are a wide variety of activities that fall under this rubric, with different objectives, different traditions and educational patterns and different needs for communication. Part of the problem and confusion surrounding discussions of the various media and their roles arises from a confusion in the types of user communities to be served and in their needs.

There is at present some difference of informed opinion as to the organic relationship between science and technology. The commonly held image that there is a continuum, with basic research at one end, and development at the other end of the spectrum, with a moving belt transmitting the results along the parts of the spectrum in a continuous way, has been challenged by a number of authors. Thus William J. Price, technical director of the Air Force Office of Scientific Research and Chalmers Sherwin, formerly of DoD, see the two streams of science and technology as generally independent: "technology usually feeds upon technology and phenomena-oriented science usually feeds upon phenomena-oriented science" (W. J. Price, 1967). Actually the channels of communication are real, between the two areas, but they are subtle and require constant cultivation.

Thus the technologist, or R and D practitioner, has widely different needs from the "frontier" research worker, and his actions and social patterns should not be expected to fit into the same mold. A recent study by Marquis and Allen (1966) has pointed out that:

> Far more is known about the flow of information among scientists than among technologists. From the knowledge that is available, however, we are led to conclude that the communication patterns in the two areas of activity are not only largely independent of one another, but qualitatively different in their nature. This difference is reflected most clearly in the mechanisms by which information is diffused within the two sets of practitioners.

A number of other authors seem to confirm Herner's 1954 relevant observation that "pure scientists are literature dependent while applied scientists are colleague dependent in seeking information".

For the technologist, the strong traditional motivation of the scientist "to publish" is often lacking. As has been pointed out by Price (1965a) from his studies of the history of technology, "one might even conjecture that the traditional motivation of the technologist is *not* to publish, but to produce his artifact or process without disclosing material that may be helpful to his peers and competitors before his claim to the private property of the technological advance can be established."

The Patent system, with its granting of exclusive rights over technical inventions, is an exchange of this right in return for the placement of the knowledge gained into the public domain through the publication of disclosure, not unlike the disclosure-recognition bargain described above for the scientific community in the form of publication.

More basically there is a difference in the role of the literature itself. As Price has pointed out (1965b):

> although there is a great deal of technological literature it does not in general cumulate by knitting in the same way as does science. Papers are not the end-product of technology, but only an epiphenomenon. There is, however, an end-product of technology in the form of a machine, a drug, a process or an artifact, and there cumulates not a literature but rather a state of the art . . . the delicate symbiosis that keeps the growth (of science and technology) in step appears to derive from the educational process that supplies scientists with a feeling for the ambient technology, and technologists with a feeling for the ambient science of their student days. Fortunately most researchers are young, so that science has available technology that is little more than a decade old, and, conversely technology can operate with quite recent science.

In any event, the technologist is anxious to capitalize on the technical advances of others and is an avid follower of technical publications and data compilations.

This distinction just made between the communication practices and objectives of "scientists" vis-à-vis "engineers" is itself subject to change. Thus the technology of today is strongly science-oriented and therefore we find that members of the different communities often find themselves in roles that require communication with other parts of the R and D spectrum. It is true that early developments in technology were often oriented around understanding of mechanics and form and were surprisingly science independent. But, as Brooks (1967) puts it,

> it seems clear today, however, that a new pattern is emerging in which the 'general powers of thought' are replacing the 'special powers of understanding of form' as primary generators of industrial innovation.

Nevertheless

> we are still not much nearer to understanding the means of measuring science in industry than to note that although some 70% of all scientists are in industry, they produce only 2% of the scientific papers and only 33% of the technical papers. Thus it would appear that the literature is no fair guide to the activities of those scientists who work in industry. In science the research paper is the product itself, but in industry it is only an epiphenomenon (Price 1965b).

Similarly, the sciences have become technology-oriented, intimately using the tools and devices supplied by their engineering colleagues and not always indifferent to the pragmatic applications stemming from their own basic advances.

In considering the realities of scientific and technical communication one must ever keep in mind that one is dealing with a social system in operation.* The individual scientist or engineer is a performing investigator within an institution and also a part of a wider national and international community. His opportunities for communication are manifold and the trade-offs and relative strengths and weaknesses of the different media are apparent to him and allow a certain flexibility. Thus Garvey and Griffith (1967a) point out that:

> In their efforts to establish and maintain contact with current work, scientists are continually on the alert for, or actively seeking, scientific or technical information relevant to their ongoing or planned work. Further, they watch closely the performance of the system as it operates to disseminate, display, and store the fruits of their own scientific efforts. If no appropriate channel exists, the producers, or the consumers, of information create new channels or modify old ones in an attempt to improve the performance of the system.

But the scientific worker is also constrained by tradition and the rigors and formalities of his environment to follow certain normative patterns. Thus, as has been pointed out by Brooks (1967):

> Although scientists like to emphasize that fundamental research is 'free', it is actually, in another sense, a highly disciplined activity. The discipline is provided

*For a scholarly treatment of the sociology of science cf. Barber and Hirsch's book of that title (1962) and Hagstrom (1965).

> by the 'scientific community' to which the researcher is related. His choice of problem and direction is heavily conditioned by the social sanctions of this community, the requirements of originality, and scrupulous reference to related and contributing work of others. . . . Although scientists are strongly self-motivated, they are also sensitive to their audience. The audience of the academic scientist is the worldwide community of his professional colleagues or peers in his own speciality, communicating through the official scientific literature, through scientific meetings, through 'invisible colleges' of preprint circulation and correspondence, and through personal contact. To the scientist in a mission-oriented organization, his audience is mixed. It consists partly of his professional community, but also to a great extent of the colleagues and superiors within his own organization.

The social role of the scientific literature in establishing priority of effort is basic to the forcing function behind *rapidity* of publication,* and the related "publish or perish" syndrome certainly contributes to the information "glut". More significantly, the tradition of publication as a way of keeping the investigator "honest" by exposing his work to the scrutiny of review and objective criticism has been emphasized as a basic ingredient contributing to the "integrity of science". This is especially significant in light of the increased business (proprietary) and national security (classified material) involvements of scientific research. Greenberg (1968), in his recent illuminating analysis of *The Politics of Pure Science*, has compared this "intricate process of certification" to the western process of jurisprudence, and finds it primarily responsible for the ways in which the institutions of science operate to advance truth, weed out error, honor the worthy and reject the crackpots!

Another phenomenon, that the business aspect of modern science and technology has engendered, involves the attitude of the engineer and research investigator to the tools and elements of his work environment. Rather than the traditional attitude of the craftsman that these should be his own property, the researcher has come to expect that his institution or sponsor should provide these amenities. Thus books, journals and information retrieval techniques are expected to be provided for him. This affluent environment has led to a lack of personal relationship with the resources and a somewhat callous lack of cost consciousness which must

*See the interesting discussions of this phenomenon in Merton (1962) and in Reif (1961). Furthermore, as has been remarked by Price (1967), "it is a most intriguing paradox that the scientist secures the maximum in private intellectual property by the device of the most open publication".

be said to contribute to the proliferation of titles, publishers and services. Unfortunately, as Parkinson has said, "when money is no object, there is economy only in thought". It is believed that a greater relationship between the information elements of support, and other facets of the investigator's resources (travel, computers, personnel assistance, etc.) would lead to an improvement in techniques and utility. The uniqueness of the market relationship in providing a "measure of effectiveness" cannot be faulted. Many services have disappeared when they have been forced to face this acid test. However, the R and D to create new information handling options may pose a problem under this "free enterprise" approach.

Ironically, however,

> technical information has no intrinsic value; its worth is determined by the social context of its use. Where there is private ownership of property, a market mechanism can set the value of goods, for one man's gain is another's loss. The gift of knowledge, however, does not diminish the giver's supply. For these reasons, measures of the value of particular items of information are very difficult to develop (Rosenbloom and Wolek, 1967).

The importance of *informal techniques* for communication of advance scientific information, as well as their related use in supplementing other phases of retrieval has been stressed in a number of recent analyses and user studies; it seems clear that this informal exchange among colleagues represents the "growing edge" of research. In describing this important concept of the exchange of scientific information through the informal media, Price (1965) has expanded on Merton's concept of "invisible colleges":

> In each really active field of science today there is now in being something which we call the 'New Invisible Colleges'—the group of everybody who is anybody in the field at that segment of the research front; an unofficial establishment based on fiercely competitive scientific excellence. They send each other duplicated preprints of papers yet to be published, and for big things they telephone and telegraph in advance. . . . By substituting the technology of transportation for that of publication they keep warm the seats of jet planes and commune with each other at small select conferences and seminars throughout the world. . . . We know little of the structure of such invisible colleges but investigations* have now clearly proved that much of the communication of research results is done long before formal publication.

---

*By Garvey and Griffiths for the American Psychological Association, for example.

In view of the dynamics of the total communication process, the interactions between elements and the substitutional nature of one element for another, a number of writers have suggested greater attention to improving techniques for assisting in the informal exchange mechanism and for better and more rapid coupling to the written media. Thus a variety of innovations have been introduced in the format of scientific meetings, in the formalization of preprint exchanges, and in the announcement of papers to be published in future issues of journals, so as to permit wider correspondence and communication, through the identification of related investigations,* etc., all to expedite the communication process.

Another important consideration in the social system concerns the "human factors" affecting the individual investigators and communicators. They are limited by the number of working hours per day (not all of which can be devoted to their subjects although, unlike other aspects of the work force, the hours required in scientific work are *not* decreasing), their rate for reading and writing, and the difficulties of assimilation and comprehension, and by their period of productivity. Their frustration is well described by their concern that they are "reading more and enjoying it less". Furthermore, they are constrained by the resources and funds available to them and their institutions to devote to the various communication elements.† Furthermore, the fact that they are the creatures of their education and behavorial acculturation must be taken into account in the design of any communication system if it is really to be used. Therefore, in both the physical and intellectual design criteria, Swanson (1966) has suggested the principle of "least action":

> It may be unrealistic to hope that any very large number of scientists can be educated to make more intensive use of present information services. We should instead assume that scientists are incurably apathetic toward them, and get on with the job of trying to change the services and not its customers. It is tempting

---

*See, for example, the new tools for exchanging information on current efforts such as the Smithsonian's Science Information Exchange, and the government-wide current effort mechanized system (after C. Sherwin).

†It is perhaps just as well that all these constraining elements of what the engineer would refer to as "negative feedback" exist, for if they did not the whole system would diverge or explode. As Price (1967) has said: "It would be no good at all if science was able to run so fast that embryonic new scientists could never catch up with the advancing front."

> to propose a kind of 'principle of least action': The design of any future information service should be predicated on the assumption that its customers will exert minimal effort in order to receive its benefits. Furthermore, they won't bother at all if the necessary minimum is higher than some fairly low threshold.

There is an even more basic issue connected with the gnawing doubt that technical people want information at all! Calvin Mooers has stated this as a law, which popularly carries his name, and which hypothesizes that: "An information retrieval system will tend not to be used whenever it is more painful and troublesome for a customer to have information than not to have it." It follows from this reasoning that many people may not want information, and will avoid using a system *precisely because* it gives information.

In view of this phenomenon it is incumbent on the overall technical community and using public to place the necessary disciplinary forces into effect which will surmount this inertia and apathy. Furthermore, it is important that information be "marketed" properly. Thus Menzel (1966) has pointed out the need for redundancy, etc.:

> Information must often be publicized repeatedly or through diverse channels before it will enter the stream of communications which will lead it to its ultimate user; and from the point of view of the consumer of information, it is frequently necessary to be exposed to the information repeatedly before it will make an impact.

This need is not always appreciated by the information "purists" who seem to be more interested in seeing to the uniqueness, or "order" in the information universe than they are in achieving dissemination, utility and progress.

In order to maintain a perspective for considering the complexity and time phasing of the various informational elements, it is important to recognize the related aspects of what we shall refer to as the "knowledge cycle".

This cycle commences with the investigator's desire to pursue a line of inquiry (experimental or theoretical) which he has more or less reason to believe productive. Later on he will go to great lengths to rationalize the logic of his reasons for choosing the particular research approach, but, at this stage we should admit that it may stem from merely a *scientific hunch* or from *intellectual curiosity*.* The investigator is in a rather unsettled state at this point; unless his research can be carried out purely with his own resources he is forced to seek sponsorship of his work on the one hand and communication with the work of those whose research bears on the matters to be investigated, on the other. He thus sets about to submit a proposal to the proper channels deemed to be interested in sponsoring the research. The resultant proposal writing has become the new artform of the post-industrial society! Guidebooks have been prepared to help the uninitiated.† Most large industrial organizations maintain staffs whose principal duties involve the preparation of such proposals, usually with strong inputs from the best minds in the organization. While these media usually are written in what might grammatically be referred to as "pre-tense", they are not without their significance from an informational standpoint. They try to exude a feeling of evolutionary confidence in building upon the previous state-of-the-art and thus usually contain a good literature search and bibliographic record which may serve as the principal literary accomplishment of the project until the much later submission of a final report or journal article. Hence their filing and retrieval within the granting organization and related governmental communities is not without utility. For this reason the government sponsors have been interested in applying greater attention to the information handling aspects of proposals, although it is expected that they will be protected as far as proprietary information is concerned.

For those who have been successful in getting sponsorship of the proposed work, their good fortune is documented in the issuance of a grant or contract award, the terms of reference for which may be promulgated in a notice of a research effort submitted to the Science Information Exchange or via the more applied current effort files of the government agencies.‡ These current effort files would presumably be consulted by

*"Scientific research usually involves the accumulation of information or the formulation of hypotheses and their verification. The design of the experiment is basically the planning of collection of data so that the desired information can be obtained with sufficient precision or that the hypothesis can be properly tested." After K. C. Peng (1967).

†See, for example, the most amusing work of Dr. Wooster (1967) entitled: "As Long as You're Up, Get Me a Grant."

‡Cf. history of the interest in such systems. (RDB project cards, DD-613's, Form 1498's, government-wide equivalents (FCST), etc.)

other related investigators or potential sponsors so that duplication could be better understood and interface efforts and cooperation better planned. (Naturally this area has been of interest to the Congress; see, for example, the Senate "Humphrey committee" reports by Messrs. Wenk and Stern, 1961.)

As the research investigation begins to achieve results (perhaps a year or more after the grant award, depending on the need for elaborate instrumentation with its associated lead time), the investigator may share his preliminary findings (Eureka!) with his colleagues through institutional colloquia, correspondence, pre-prints and letters to the Editor, etc. Of course, progress reports will be submitted to his sponsor on a more or less periodic basis.

Subsequently, the researcher may have the temerity to "give a paper" before his professional society, available to his colleagues via an abstract in their society bulletin and perhaps a pre-print made available to those in attendance. At some later time (usually several months later, from the findings of Garvey and Griffith (1965)), and usually following a final technical report to his sponsor, he will submit the material in the form of a scientific paper to a journal. After refereeing, correction and publication it is disseminated to the international community which has "current awareness" of the material. However, the attention of most people will be called to the work principally by coverage of the work in an index or abstract "secondary publication", thus leading to a further six months delay, approximately, in the cycle.

Upon subsequent refinement and critical review, the work, if it has sufficient merit, may be incorporated into a technical review article on the subject area, eventually being referenced in a monograph on the subject. Finally the work may find its way into an educational curriculum, eventually becoming a part of the "conventional wisdom" of the technical community.

The cycle just described may consume a decade and invariably sees many modifications of the results and concepts in the course of this elaborate refinement process.

In reflecting upon the dynamics of the scientific communication process, one must recognize the operability of a kind of speed-accuracy analog of the physical "uncertainty principle". Thus the review and modification process, which is basic to the self-correcting evolutionary refinement of scientific knowledge, must be accomplished through the expenditure of time. To penetrate this cycle at any intermediate point for the purpose of *using* information, must of necessity involve problems of comprehension and risks of error. Nevertheless, this process of interaction with information in the various states of refinement is basic to the progress of science and technology, and to the publicly desired acceleration of its diffusion and exploitation.

The *traditional* scientific communication system has evolved over the last few hundred years from a small number of stylized, normative, patterns:

The Scientific Meeting
The Scientific Journal
The Scientific Book
The Scientific Library

The first two media have traditionally been organized through the learned and professional societies. The monograph or book was originally published through university presses or other scholarly publishing houses, while the great libraries were traditionally associated with the universities or national cultural institutions.

In recent times many important changes in the organization of scientific institutions, and in the support and utilization factors of scientific endeavor, have caused analogous changes in scientific and technical communication patterns which have introduced many new media and have placed great pressures upon the older media. Thus, as science and technology have become economically significant elements, commercial interests have entered on the scene and are now involved in a host of enterprises which involve all phases of the communication system.

As the role of the Federal Government widens in support of the continually expanding R and D activity in the nation, it is reasonable that it should be concerned with the viability of the scientific communication channels and with the opportunities for disseminating the information resulting from its extensive R and D programs. Where the Government is the sponsor, it has come to recognize its fiduciary responsibility for the wider dissemination of the information resulting from its support of R and D activity. The various sponsoring Agencies of the Federal Government have also come to recognize the important principle that a research effort is really not completed until the conclusions, findings, and data are disseminated.* It is also coming to recognize that the research work itself will profit from, and be validated in the course of the evolutionary process of self-correction and refinement which proceeds from widespread technical community access to the results.

While the Government's objectives are not always explicitly stated, a recent Air Force "STINFO" statement has put the purpose succinctly: "To improve the exchange of scientific and technical information so that (1) the information needs of scientists, engineers, and managers are fulfilled; (2) Research, Development, Test and Engineering costs are reduced; R.D.T. and E. cycle time is reduced; and (4) R and D management is improved."

The extent to which management can be served by better information systems has been of special concern to Congressional bodies.†

Government R and D technical reports are, therefore, made available through clearinghouses and agency information systems to other contractors and to the public at large on a "need-to-know" basis and with some minor charges for handling and reproduction. The Government has also recognized a much wider responsibility for the well-being of this scientific communication "resource" and has sponsored a large number of improved tools, indexing systems, secondary journals, library innovations, and specialized information and analysis services to provide more expeditious and effective exploitation of technical information. In doing so, and in its wide use of the *technical report* format, it comes into interface with various other elements of the professional, industrial, and commercial communities.

*Dwight Gray (1962) has expanded on this belief that "dissemination of the results of experimentation is an integral part of the total research process" to the recommendation that "there should be a commensurate increase in the funding of the former with an increase in the expenditures for the latter analogous with the obvious expansion in the facilities, etc." Data on the expenditure for scientific communication are very difficult to compile and usually include only the very obvious and explicit expenses (e.g. to operate DDC, FCSTI, etc.). It is therefore difficult to judge whether this admonition has been listened to.

†See the Wenk Report (1959), Stern Report (1960), Crawford Task Force (1962), Elliott Report (1964), Weinberg Report (1963), etc. Note recent current effort renewed program (1498's).

The advent of new technology involving digital computers, efficient composing and printing innovations, widespread reprography techniquest, greater opportunities for person-to-person contacts through the use of long-distance communications, and the affluence which permits rapid long-distance travel, have greatly modified the environment for technical communication.

Furthermore, the division of labor which worked so well in the process of specialization of *investigators*, has now been applied to the information field itself and we see the proliferation of a great many middlemen who are involved in the various essential indexing, abstracting and information-analysis center activities.

Nevertheless, in spite of all these changes, it is still true that there is no royal road to learning, and the traditional functions of scholarship and its associated apparatus have by no means been substituted for by new technology. We thus see an even more important place for quality of publication, precision of writing, intellectual classification and processing of work for retrieval, and most significantly in the preparation of reviews and monographs.

We thus find the area of scientific and technical communication to be very complex and intimately related to the scientific and technical community which it serves, with its diverse social and economic goals and behavior patterns. The current situation exhibits a not always "peaceful coexistence" among the various elements; the scientific and technical journals, the technical report literature, the secondary indexing and abstracting services, and the commercial "knowledge press", all complicated by the growing apparatus of new technology and the intricacies of government and private sector relationships. Their characteristics and roles cannot be simply partitioned on the basis of function or type of literature; rather they interact and compete in various ways for the attention of authors and "users" and for the available resources. While this competition and redundancy may very well have some positive attributes, it is worthwhile to consider the problem areas in the various relationships between them and in their internal operations all of which can profoundly affect the desired success of the total information and communication system in performing its crucial functions to assist in accomplishing the goals which the nation seeks from the world of science and technology.

One of the areas that has received a great deal of attention in the information field has been the attempt to assess the "needs" of various users in performing their functions for research and development. This is no mean task since it involves considerably more than merely asking—although numerous surveys of this type have been conducted.*

For a good review of the growing body of user surveys, especially from the sociology standpoint, see Menzel (1966). National Science Foundation (Office of Science Information Service) has provided funds for several professional societies to introspectively study their members' needs. The Department of Defense has had studies of the information requirements of both its in-house research and development staffs (Auerbach Study, 1965), and of its contractor personnel (North American Aviation, 1966). These studies tend to confirm the importance of internal information elements, especially the assistance that colleagues offer for referrals and state-of-the-art assessments. Another result of these studies is the identification of the abysmal ignorance and apathy on the part of many workers towards the existing information tools. Some method of updating workers into a more comfortable and knowledgeable relationship with these elements is clearly necessary.

For a critical commentary on the findings of these user surveys which show a heavy emphasis on informal contacts with colleagues for information, we agree with Dr. Y. S. Touloukian of Purdue, who at the AIAA conference on October 23, 1967 said:

> Among all of these swift currents, in recent years certain sources have revealed that studies of *communication habits* among engineers and scientists show that their prime mode of communication is verbal and personal and not the printed document.
>
> The promotion of such a notion has indeed had beneficial results to engineers and scientists as it has stimulated among a number of governmental agencies more generous funding for travel to meetings and conferences. While I can hardly object to such a policy, I must state without reservation that the so-called 'factual finding' stated above is without sound basis, nor is it true in principle. The fact that an engineer finds it more convenient to query his officemate or the

---

*As has been pointed out by Eugene Wall (1967): "There are undoubtedly cultural, sociological, political, psychological, and other similar, often nonobjective, obstacles to the satisfaction of the true needs of users. In order to detect some of these obstacles (as well as more objective problems in the economic and technical areas), scientific disciplines should be encouraged to undertake critical self-examinations."

fellow down the hall for his information needs does not infer that he gets the information he seeks nor that he obtains the correct information. Indeed, one may ask where his officemate obtains his information. A truer diagnosis of the situation would be that our engineer finds it difficult to have quick access to the proper information sources or more often he is not aware of the availability of these sources. His action rather represents a reaction to a faulty communication system resorted to through desperation. The logical answer to the problem is to make the information available in a form that is readily accessible and in scope adequate for his needs.

User needs can be categorized functionally. Thus, after Voight (1961), there are the principal needs of:

(a) current awareness,
(b) retrospective search for specifics,
(c) exhaustive search of literature.

Loosjes (1967) gives the five phases of a complete retrospective literature survey:

1. Definition and limitation of subject matter.
2. Search strategy scheme.
3. Looking up titles.
4. Retrieval of the literature itself.
5. Evaluation.

These are all best left to the research worker himself.

A more detailed breakdown of needs has been given by Conrad (1967):

Help in browsing.
Specific answers.
Knowledge of prior work on problem.
Knowledge of what has *not* been done.
Knowledge of both fruitful and unfruitful findings.
Who is working in field.
Applications.
Costs.
Manufacturing facilities available.
Status of competitive situation.
Existing and potential patent constraints.

A still more empirical breakdown of the needs of research workers has been given by Rosenbloom and Wolek (1967), who divide the acquisition modalities into two categories depending upon their *competence-oriented* or *problem-oriented* nature. For the former, sources outside the corporation are used substantially more often.

> As would be expected, documentary sources are most common in these circumstances; among scientists they are almost entirely made up of articles and books which constitute the 'professional' literature; among engineers the use of professional sources is matched by the use of information from trade magazines, commercial catalogues and technical publications, and technical reports of other organizations. . . . Among *scientists*, sources within their own corporation provide information in only 1/3 of the instances, as opposed to the strong majority (50–70%) of the instances reported by most groups of *engineers*.

They further distinguish between the professional-oriented activities vs the operational-oriented activities of personnel. In any case, it is information transfer that is significant not just information retrieval. "In the transfer of technical information in industrial laboratories, instances of information looking for the man seem to be nearly as frequent as the efforts of men seeking information."

One must also recognize the somewhat circular relationship between needs and new techniques. Thus, as John R. Pierce (1968) has pointed out in considering the history of the now essential telephone and the automobile, there was no pressing "need" for either at the time of their creation, from which he concludes that "*needs* come into being and grow as new *means* come into being."

More basically, one must consider the communication needs from the standpoint of those factors that significantly enhance the actual progress and accomplishments of the scientific and engineering processes, and are not merely ephemeral. Historical and other studies of "events" that significantly contributed to such progress are getting more popular in connection with the "Science of Science".

A recent study reported by the National Academy of Sciences (1966) identified these parameters as being significant for stimulation of the performing engineers:

1. Flexibility for the individual investigators to make major changes in direction and goals was frequently required.

2. Close and frequent communications between organizationally independent groups were often essential.
3. Key individuals played essential roles in bridging the geographical, organizational, and functional barriers between groups.
4. The recognition of an important need was most frequently the principal factor in stimulating research-engineering interactions.
5. Often technical approaches were available and lay dormant for some time before their pertinence to a specific need was recognized.

CHAPTER 3

The Primary Scientific Literature

A. SCIENTIFIC JOURNALS

Kronick (1962) has traced the development of the "scientific journal" and has shown that early periodicals could be considered printed letters addressed to a wider audience than could be reached by the written letter. As has been noted in Dr. Weisman's (1968) recent treatise on Technical Correspondence,

> correspondence has a long, venerable tradition in science and technology. Letters were used by the earliest scientists to exchange information, to report activities and discoveries, to try out ideas, and to receive critical evaluation of experimental results. Frequently (as in the case of Galileo and Kepler), letters were used to justify and defend scientific works. A letter from one scientist was circulated among colleagues at great distances much in the same way as reprints are circulated today.

While technical correspondence is still utilized today more than may be superficially thought to be the case, the formalization and business aspects of technical investigations require a more significant and wider broadcasting of results. It is interesting to note that attempts to revive this original phenomenon, by formalizing the exchange of correspondence to a readership of interested parties through the so-called "information exchange groups" in biology, high energy physics, etc., has been resisted by the journal "establishment" who see great danger to their journals and to science as well by such unreviewed proliferation of preprints and miscellany. This will be discussed further below.

With the growth of scientific activity, the grouping of scientists together in "societies" allowed for discussion of scientific results at "meetings". The "Proceedings" of these scientific societies eventually evolved into journals of those societies to serve their membership's communication needs. One gets the distinct impression that the early driving force for this communication innovation was simply the pleasure of doing good work and the desire to tell others of it. It is hoped that the pragmatism of modern science and technology does not completely stamp out this attitude which still persists in many workers today.

While in principle most journals still attempt to emulate the traditional format, the enormous expansion of science and the proliferation of new disciplines and hyphenated interfaces between disciplines, have greatly strained the society–publication relationship. This has caused two major stresses in the journal field. On the one hand the contents of society journals are of great significance considerably beyond the boundaries of their society memberships, and the proper apportionment of journal costs and the economic viability of the journal enterprise present a grave issue. Secondly, the likelihood of a one-to-one relationship between the contents of any society journal and the needs of any using community has been greatly modified by the patterns of intensive specialization which are so much a part of the scientific scene. "This occurs to the detriment of the recipients who must allocate valuable reading time to the relatively unrewarding pursuit of scanning much that is irrelevant in order to read a few things of importance" (Swanson, 1966).

Techniques for coping with this problem include the wide use of secondary indexes and in the modern eclecticism known as selective dissemination of information (SDI), both of which will be described in detail in subsequent sections.

Operationally, an NSF (OSIS) study (1964) found that:

> scientific periodicals serve two major roles: a vehicle for the communication of new discoveries and ideas, and a repository of knowledge. The latter role is a relatively new one. By forcing the periodical to play a double role . . . it is possible that we have forced upon it an impossible task which is now threatening it with a complete breakdown.

Journals, as communication media, have become national and international resources and appear in a wide spectrum of types and support mechanisms. As science and technology have become business-oriented, the commercial publishers have entered in force into the periodical communication media, just as they had done earlier in assuming a dominant role in the monograph and textbook field.

Meanwhile, the traditional scientific literature, published by learned and professional societies, has continued to persevere and indeed to expand. Although greatly strained by the pressures of the growth and specialization of science and technology, this "canonical"* literature has continued to emphasize the same distinguished characteristics of quality, archival status, and open availability that have continuously provided basic contributions to the strength and integrity of scientific progress. The strong force which such societies exert over the literature provides quality control and assures coherence and relevance, principally through the discipline and assistance which their memberships provide in generous support of the operations through editors, reviewers and membership feedback via comments.

In spite of new techniques for improving publishing speed and efficiency, massive doses of volunteer labor, and some limited government support through authorization of page charges, most societies find their publishing activities to be a severe and noticeable strain on their economic viability. This situation is exacerbated by the widespread violations of copyright protection through new reprographic innovations as well as the challenged "tax free" use of technical advertising revenues.

Commercial and trade journals have become an increasingly significant element on the "periodical" literature scene. Modeled after the "society journals", making wide use of prestige editors and editorial boards and, in many cases, stressing the international nature of scientific publication (filling a vacuum due to the general absence of operational international publishing professional societies), these publications additionally supply journalistic and entrepreneurial professionalism. They face the rigors of the supply-and-demand market-place, and take liberal advantage of the important element of technical advertising and its need of a vehicle to reach an audience† as well as the pre-conditioned respect for publications

*This term was suggested to us by Dr. J. Weyl (Executive Secretary of the National Academy of Sciences Committee on Scientific and Technical Communication (SATCOM)). We believe that it evokes the proper admixture of the standard, traditional, and authoritarian characteristics of this literature.

†Galbraith (1967), in his interesting book *The New Industrial State*, emphasizes this requirement of the industrial "technostructure" to stimulate demand and to interface with what he refers to as the "scientific and educational establishment" to achieve a symbiosis. Advertising is the preferred medium for this communication and the requirement of an effective vehicle has been responsible for the wide-spread format of so-called "controlled circulation" type publications which are sent gratis to a selected audience.

which have caused institutional libraries to become universal collectors of almost any and all publications offered.

Thus we see that there is today not only a large and dynamic population but a variety of journal types which should be distinguished, since their different characteristics are significant to our consideration. We give here a brief taxonomy of the journal literature:

1. The standard type journal represents the archival publication of a professional society. Many of these journals still reflect the format of a "Proceedings" of a learned society. The society is very much in charge of the personnel and policy of the journal and the officers are subject to the elections of society members. The budgets of such journals are a part of the society economics. Editorial review and referees are chosen from professional society membership, etc.

2. Professional societies have found it necessary to develop additional periodicals to keep pace with the communication requirements of their members. This has taken several directions; some have recognized that there is a need for a more rapid and limited format for early description of findings but which still attempts to retain the quality control aspects of review. Some call this material "letters"; societies find it possible to publish this material in as little as one-third the time necessary for full articles. This type of material is very similar to the "notices" and "reports" which have appeared for some time in rapid and interdisciplinary scientific periodicals such as *Science* and *Nature*.

Other journals have successfully innovated along the lines of more applied subjects with somewhat less formal pretensions of archival significance.

The fact of increasing specialization of science has forced societies to a greater and greater narrowing of the field of coverage of their journals through introducing new series within the publication program "family". Thus, for example, the Institute of Electronics and Electrical Engineering (IEEE) now publishes about forty different titles in its "transactions" series.

In addition, societies have introduced new magazines which serve to focus on the identity of interests and professional affiliations of their members and which supply general and tutorial materials. These often include a considerable amount of technical advertising which may be of significant browsing value as well as in enhancing the society income to assist in their other less viable publications.

3. A number of commercial publishers have been extremely successful in organizing journals which in many ways resemble those of professional societies. They are particularly successful in new budding branches of older fields and in areas where an international community of interest exists; in both cases the flexibility and lack of inertia of their free enterprise operation has been a significant element in their early staking of a claim. These journals seem to be growing in significance and numbers to the point where they make up as much as a quarter of the titles in certain technical areas. While the commercially sponsored journals do not always have the same professional overview as does the learned society journals (see below, re quality considerations) there are nevertheless some clear incentives to good publishing practice, especially those relating to user acceptance. Thus these publishers seek to attract professional quality editors and usually give them complete policy control over the journal.

4. A number of periodicals are published under government aegis. This is, of course, a necessity in the case of security classified materials but it also extends to areas in which government laboratories represent an important element in the information generating field, e.g. The National Bureau of Standards' Research Journals.

5. Likewise, a fairly large number of corporate institutions publish journals, some of them dealing with basic contributions to scientific knowledge (e.g. the *Bell Systems Journal*), others dealing more with applied or institutional programs bordering on the "house organ" style. The principal objective of such journals would appear to be more the recognition and approbational encouragement offered to employees and to the corporate image than to the diffusion of knowledge, however, it is not unlikely that the latter purpose is served as well in the process.

6. Commercially operated trade journals are an especially significant element in satisfying the needs of the engineering community. These are strong in the state-of-the-art review and tutorial subject areas and carry a generous quantity of technically informative advertising.

A communication area wherein private publishers have been most successful, has been that which we would characterize as "technical intelligence" for industry. Trade publications are especially competent in ferreting out information and providing the type of leads which are invaluable to industrial R and D people to assist them in their interactions with each other and with the Government.* There is an important additional role for the profession of journalism in providing the various specially packaged information products, which assist R and D practitioners to locate products, instruments, vendors, etc., with precision and detailed awareness of capabilities. We do not think that this same enterprising spirit and journalistic know-how could be provided through professional society undertakings or by government groups; furthermore, we believe that the opportunity for critical commentary which they often provide, is a necessary check and balance which provides a positive influence on the integrity of the communication system. (In this connection their larger readership gives weight to this element; e.g. *Aviation Week* is read by 330,000 people.)

7. The need for review materials in science, and the need for interdisciplinary dialogues and the growing significance of science policy matters, together with the greater affluence of the business of science has led to a large number of subscription and "controlled circulation" type journals which have found an important place in the competition for the time and attention of the scientific and technical community.

Some statistics of the journal literature are useful to properly appreciate the magnitude of the issues and to acquire some perspective. Since science "began", it is estimated that about 10 million scientific "papers" have been published. We are adding to that number at the rate of about 6% annually (i.e. a doubling time of about 12 years). These 600,000 new

*Note, however, that a distinction should be made between "scientific information" and "scientific news" (see also McKenzie in Weil (1954) and his related category of "science in the making"). This is related to the archival vs. current awareness function distinction.

"To contain scientific information, a scientific paper must be written in such a way that an able research worker, specialized in the same branch of science, will be able from directions given in the paper and from these only, to reproduce the experiments and obtain the results described, with errors, smaller than or equal to the margin stated by the author; . . . papers and letters need not correspond to this definition; since the statements they contain cannot be verified by independent workers, we would prefer to call them Scientific News: they play in the scientific world the same part as ordinary news does in the world at large. For instance, let us suppose that somebody writes that he has been able to synthetize (*sic*) diamond: this is scientific news. If his paper describes correctly and completely the operations by which this synthesis was made, then we are in the presence of new scientific information" Boutry (1959).

papers a year appear in some 35,000 current technical publications, over 6000 of them in the United States.* Any consideration which assigns equal weight to the significance of each of these many journals would, of course, be erroneous. Some students of "citation indexing" who have made pioneering studies of "literature coupling" and "paper networks" have purported to show that most of the original research in the United States is published by a "hard core" of about 400 scientific journals. Other evidence points to greater scattering of information in the periodic literature. Thus,

> although perhaps half of the articles on a particular subject are concentrated in a dozen or so journals, the rest are scattered widely in many hundreds of journals, few of which could have been predicted beforehand as likely locations. In short, a substantial fraction of scientific literature on any one subject is so scattered as to defy the efforts of any one individual, even with the advantage of reasonably good bibliographic resources, to bring it all together. *The impression of many scientists that they are 'keeping up' by reading a half dozen journals regularly may be illusory* (Swanson, 1966).

Delving further into the demography of the journal literature, D. de B. Beaver (1964) has observed that the number of authors of n papers is inversely proportional to n^2, thus the distribution of papers amongst authors exhibits great inequality; as few as 10% of the authors are responsible for 90% of the papers. As regards citations, papers cited are randomly distributed through the entire literature of a field except for a greater citation frequency for papers less than 10 years old superimposed upon the random pattern. In addition, on journals themselves, their statistical study of journals shows how papers are distributed amongst journals, and reveals an inequality almost as great as that for papers and authors. Moreover, it results in the striking conclusion that the large majority of scientific journals are highly ephemeral and therefore contain but a small fraction of the totality of scientific literature.

While the number of journals in existence is exceedingly large (~ 35,000), it is not as large as some modern estimates based on extrapolations from growth alone would have estimated (~ 100,000) because of the phenomenon of "journal mortality". Beaver found that half of all the journals ever founded lasted less than 6 years; in fact 30% last 2 years or less.

*For derivation and extension of these statistics cf. Kronick (1962), Gottchalk and Desmond (1963) and Price, D. J. (1963).

More than 75% (by volume) of the total scientific literature is contained in the oldest 25% of the journal population, but to make the problem more complicated, young journals contain a substantial fraction of the scientific literature of each new year and there is, of course, no way of identifying their life expectancy.

Most of the estimates of journal titles have one thing in common—they are arrived at by analyzing lists of serial titles. A more pragmatic survey was made by the British National Lending Library (NLL) who took on the Herculean task of actually seeking to acquire *all* legitimate scientific serials extant. Based on their actual experience with phantom journals and other non-starters, they estimate the total journal population as "only" 26,000 titles, and they have the material to prove it! (Barr, 1967).

Journal statistics are, of course, only one side of the question. Methods of use and reading habits of users are the other side, and are much more difficult to quantify. Some data suggest that about half the articles in core journals in the field of psychology (which contain some half of the total literature and most probably the most significant half at that), will be immediately read, in this specialized sense of the word, by only 1% or less of a random sample of psychologists, and no research report is likely to be read by more than about 7%. The data clearly indicated that the immediate audience for most articles is of an extremely restricted size. The number of current readers increases very slightly as the number of persons seeing the issue increases, but even the most popular issues (seen by 20% of the sample) contained a sizable percentage of articles that were examined by 1% or less of the sample.

> The articles read by 1% or less of the persons sampled—that is, half the articles appearing in "core" journals—would by extrapolation to the population, have a total of about 200 readers. Thus the immediate dissemination through journals is well within the range of that for some of the other forms (preprints, reprints, technical reports), which, because they are sent to interested persons, are likely to have high rates of use (Ronco, 1965).

The small actual readership of the scientific literature may have a more profound explanation than the sheer statistics may convey. Thus Price advances the belief that the primary social origin of the scientific paper is the desire of the scientist to stake and record his claim. Only incidentally, he complains, does the paper serve as a carrier of information, an announcement of new knowledge for the good of the world, or a gift of free advantage to one's competitors. Claims to scientific property, he argues, are vital to the makeup of the scientist and his institutions; scientists therefore "have a strong urge to write papers, but only a relatively mild one to read them" (Price and de B. Beaver, 1964).

While mention has been made of the "core" journals and the significant observation that they contain much of the literature cited by other authors (see, for example, the work of Garfield and that of Price in illuminating analytical studies based on the *Science Citation Index*), it should be pointed out that the literature "distribution tail", which contains a great many of the minor journals is still quite important. Even the redundancy of literature, itself, is an important element, in terms of the ways in which interdisciplinary information diffuses and in the ancillary information concerning important but perhaps non-central information about a work, including instrumentation, etc. This phenomenon is perhaps more significant in applied fields where redundancy is expressly encouraged and given the more pragmatic term, "repackaging". The goal of scientific communication should not be based on uniqueness and spartan efficiency of publication but rather in terms of the diffusion of knowledge and in the feasibility of its exploitation. As President Johnson has aptly put it: "our generation will be judged not so much on the literature we accumulate but rather on the rapidity and effectiveness with which we apply the knowledge contained in it."

An interesting para-phenomenon related to the proliferation of literature is the increased number of authors associated with an individual research work, which is of course more prevalent in those areas of "big science" where a team effort is necessarily involved:

> In Physics, of the papers published during the decade of the 1920's, 75% were by single authors; in the next decade, 56%; then, in the 40's, 50% and finally, in the 1950's single-authored papers declined to 39%. A similar pattern in biology begins later and develops at a slower rate (Merton, 1965).

Sociologist Merton (1968) has studied this effect, including the fascinating corollary which he calls the Mathew principle; this principle states that unto those who have, shall be given more—or in the field of literature it confirms that the prolific authors tend to dominate the publication field, especially because of the team concept where the team captain is associated with all the works and is usually credited as senior-author.

There is a tendency to treat the problems of scientific and technological communication as general in nature, with analogous problems for all disciplines. To a large extent this is true and probably reflects the "universality of science", however, there is no question that there are special problems which characterize and differentiate the techniques and procedures of the various science and technology divisions. Thus W. Koch, Director of the American Institute of Physics, has pointed out that:

> Physicists understand that their science is an intellectual effort that is organized and pursued in a manner different from that in which chemistry or biology, for example, are pursued. Physics has a conceptual orientation rather than the structural orientation of chemistry or the functional orientation of biology. Therefore an information system for physicists will have a different character from one for chemists, biologists, geologists, etc." (Koch, 1968).

It is also true that some disciplines have become alert to the information *flood* earlier than others (cf. Licklider, 1966) and have sought the *higher ground* of better information retrieval practices to avoid complete inundation. It is interesting to note in this regard, that the field of chemistry which, as Westheimer has pointed out in his NAS chemistry state-of-the-art report, is primarily supported from private funds, has taken earlier and more comprehensive steps to improve the information handling techniques than many other areas far better supported by public funds! (See, for example, American Chemical Society and Institute for Scientific Information efforts.)

The role of professional societies in handling the communication aspects of their disciplines is being strained by the proliferation of journals (caused by interdisciplinary, international and proprietary pressures) as well as by economic problems. Nevertheless, we believe that they continue to offer the greatest force for good in this area and must continue to play the intellectually civilizing role that we have come to expect from them. In spite of the growth of competing activities, they still publish the prestige core journals which contain the most valuable and significant research papers and they are of necessity closely coupled to their users, both authors and readers, in a direct and most responsive manner. In the U.S. the principal governmental groups, with support responsibility in this area, have continued to recognize this significance of professional societies and have been advised to respect and nourish this element. Thus the National

Science Foundation—Office of Science Information Service has traditionally applied its resources through such societies (AIP, ACS, etc.) as it has been advised to do by NAS–SATCOM and its Science Information Council. However, within other phases of government (FCST–COSATI), DoD, etc., there has not always been this enlightened attitude. Vigilance is required on the part of the scientific and engineering community, to see to it that the government operators of information systems do not act in harmful or indifferent ways to the detriment of the valuable contributions that learned societies and their communication channels may provide.

B. QUALITY OF THE TECHNICAL LITERATURE

The importance of quality in the technical literature cannot be overstressed. However, it must be said that this is an elusive and sophisticated concept which is not at all easy to quantify. It is in the very nature of scientific progress that old and cherished concepts are overthrown for new ones, and the "conventional wisdom" itself is no guarantee of truth. Scientific endeavors do not necessarily lend themselves to the traditional democratic principles—in science one must ever keep in mind that a belief by a majority that something is so does not make it a fact. In fact, for science, "truth" is a relative concept evolving along with our greater understanding of phenomena and the improving accuracy of our measuring instruments.*

From experience, we have come to depend upon certain important quality control mechanisms in the scientific literature:

Technically sound qualifications and credentials of the author, including technical objectivity in his purposes.
Open publication for all to see and criticize.
Explicit statements regarding the experimental conditions, results and the precedent works which were assumed or included.

*". . . What is truth?" After nineteen centuries we are still getting new answers to this question posed by Pontius Pilate. The reason for this lack of a definition is clear enough. The universe is so complex that even the widest-ranging vision is at best partial and tentative. Each generation sets itself the task of reevaluation and reinterpretation, and so the quest for truth is unending (H. Eyring, AAAS Presidential address).

A posteriori criticism and comment on the work by those who see relationships with their work, or who have been able to confirm or possibly to take exception to the findings.

The journals of professional societies have built up an eminent reputation on all these counts and rightly regard them with cherished respect, and look with suspicion on any tendencies to short circuit or degrade them. As was discussed in the preceding chapter, and as is well expressed by Alvin Weinberg (1967):

> . . . the scientific community has evolved an empirical method for establishing scientific priorities—for deciding what is important in science and what is unimportant. This is the scientific literature. The process of self-criticism, which is integral to the literature of science, is one of the most characteristic features of science. Nonsense is weeded out and held up to ridicule in the literature, whereas what is worthwhile receives much sympathetic attention. This process of self-criticism embodied in the literature, though implicit is nonetheless real and highly significant. The existence of a *healthy, viable, refereed* scientific literature in itself helps assure society that the science it supports is valid and deserving of support. This is a most important, though little recognized, social function of the scientific literature.

Nevertheless, as has been stated, the issue of quality is at best, subjective; journals differ in the thoroughness of their use of expert review and the review process itself is quite variable.* The tradition of anonymity in the review process has its obvious drawbacks, although it does not prevent an exchange between the reviewer and the challenged author; often the issues in contention are ajudicated by a third party serving as a "referee". (There have been various suggestions to improve objectivity in the referee relationship such as making the author's identity unknown to the reviewer, etc.) Many outstanding scientists have been notoriously hard on reviewers.† Take, for example, the well-known case of W. W. Coblentz at the National Bureau of Standards, whose careful works on infrared spectroscopy are still referenced as significant, some 50 years after their day. Dr. Coblentz maintained that he alone was responsible for the quality of his work and refused to accept the criticism of his colleagues or journal editors and reviewers— "people who knew less about the work than he

*For an interesting review of journal editors' attitudes about the reviewing process cf. the survey and analysis in Passman *et al.* (1968).

†See, for example, the tongue in cheek discussion of "A Rational Scale of Critical Editorial Evaluation" (Frechette and Condit, 1967).

did". In his autobiography, *From the Life of a Researcher* (1945), he advocated a system of scientific publishing which would throw the burden for quality completely on the author, whose reputation alone was at stake. In essence this is the defense offered by the Information Exchange Groups and some new unrefereed rapid journals.

In view of the spreading concern over the "information flood" there has been a turning to the *reviewing process* as the answer to coping with the pressures on the part of authors, who must publish or perish, and on commercial publishers who need new titles for their business.* We expect that this issue will always be with us and will require diligence from all concerned with the information process.

The important winnowing techniques offered by state-of-the-art reviews and monographs, and the rise of the specialized information and analysis centers for selectivity and reviews will be discussed below. This is especially important because, as has been noted by Swanson (1966), "knowledge that articles will be judged by their colleagues now constitutes some incentive to authors both to write well and to write only when there is really something to say. Better methods of quality indentification within the literature should strengthen this incentive."

C. ECONOMICS OF PUBLICATION

Accepting the scientific journal literature as a highly significant element in the scientific communication system, we should naturally be concerned with its intellectual health and well-being. It is not enough to pay it lip service, but it is necessary to see that it can continue to provide the important functions upon which scientific progress depends. One of the principal threats to its existence is in terms of its economic viability.

*Unfortunately, the resourcefulness of those who seek to publish cannot be ignored. Thus, as is pointed out by Garvey (1964), in discussing the dynamics of the scientific communication process:

> "Changes or growth occurring in one element affect in some way and to some extent, the operation of other elements in the system. For example, increasing the rejection rates of manuscripts submitted to journals almost automatically increases the birth rate of new journals. Unfortunately, decreasing the rejection rates, which usually results in increased publication lag, does not reduce the birth rate of new journals, but it probably does decrease the current use of journals and certainly increases the number of preprint exchange groups."

An important aspect of the economic vitality of the journal and the review literature as well, concerns the manner in which this intellectual property is to be protected from "unfair practices". The copyright law, with its potential penalties against infringement, has been the protective technique since the earliest days of publishing. However, new and widespread methods of reprography and modern techniques of mechanization of information retrieval systems have so perturbed the previous control system as to require a new look at the issues.*

With regard to the journal literature, the great multiplicity of specialist journals and the significant cost of subscribing to them (many monthly specialized journals now charge over $50 per year for a subscription) together with the new ease of copying, have fostered the following by-pass techniques in their exploitation. Individuals have become dependent upon corporate, regional and national libraries for access to the journal literature.† It is usually the case for a scientist to retain a small number of memberships in professional societies with which he feels a technical kinship and to subscribe to only their basic communication journal.

Institutions serving thousands of engineers and scientists often subscribe to only single copies of important journals.‡ The contents of these journals are made known to the staff either through library scanning by individuals, or commonly through rapid circulation of internal institutional cumulations of article titles or, alternatively, via a subscription to secondary title

*A revision of the copyright law has passed the House in the 90th Congress but is still being considered in the Senate. Most likely, however, the revision will temporize with this issue and postpone attempting to solve this problem until a national commission can consider the issues. For a good review of the problems in this complex legal–technical–economic domain see Marke (1967) and Sophar (1967).

†This is part of the general increased reliance on corporate resources for material necessary for the performance of one's work, a trend away from the practice of a skilled worker's pride in personal possession of his work tools. The trials and tribulations of being dependent upon corporate subscriptions leaves much to be desired; e.g. in the time consuming routing of library journals, and in the requirement of returning materials as needed by others rather than having them permanently accessible to the individual who would benefit from their entering into his personal retrieval system. New opportunities to enhance one's individual reference collection, based on "SDI" techniques, will be discussed later on.

‡It is true that publishers often charge a differentially higher subscription price to libraries and other multiple users, however, it is certainly not adequate to substitute for any appreciable number of individual subscribers.

references such as ISI's *Current Contents* (with separate series for Physical, Chemical and Biological subjects). Orders for articles deemed to be of interest, are then placed with the libraries,* who commonly have adjacent reproduction machines to perform a vast copying service, without either any permission from the copyright owners or any attempt to make fair recompense.† Furthermore, upon the further diffusion of awareness of articles of interest through secondary abstracting and indexing services (e.g. Index Medicus, etc.) requests can be made to central libraries such as the National Library for Medicine‡ and the British National "Lending" Library,§ where copies are made from their vast single-copy holdings. Only the Library of Congress, because of its responsibility as the national copyright repository, appears to pay any heed at all to the amenities of copyright protection in photocopying.

From the standpoint of the libraries, they seem to have no alternative but to honor their users' requests for photocopies and they rationalize their actions by referring to the so-called "fair use" doctrine which has come to consider single-copy production for scholarly use as proper.¶ However, it would appear to us that a distinction must be made, at the very least, between the actions of an institutional library serving its internal staff, and the large "wholesaling" national libraries who can and do

*Subscribers to *Current Contents* are also offered a "tear sheet" service from ISI's holdings of the primary materials (OATS).

†In a recent survey, Mr. G. Sophar of CICP (1967) found that this copying already amounts to at *least a billion pages* per year and is growing!

‡The National Library for Medicine in Bethesda, Maryland, currently reproduces from journal hard copy but plans to set up a microfilm computer system for more efficient reproduction. Even when notified by journal editors to cease and desist from these practices, the Library has reputedly indicated that it will stop only when and if required to do so. Legal action in this direction has now been taken by the biological journal publisher, Williams and Wilkins; it will be interesting to see how the courts rule on the "fair use" principle.

§The National "Lending" Library (Yorkshire Spa, England) (Urquhart, D. J., 1959, 1962) has a collection of 26,000 journal titles from which over 2000 reproduction requests are serviced daily! The massive use of reprography which they practice makes their title involving lending a euphemism. They are now seeking to expand their user service so as to provide worldwide rapid reproduction services!

¶See the survey by CICP (Sophar, 1967) of the rationalization by librarians on this point. The survey responses seem to indicate clearly that the librarian "either firmly believes, chooses to believe, or chooses to take the position" that the making of a single copy of any part of a copyright work is fair use.

serve the entire world population. For them to base indiscriminate copying services to all and sundry requesters on this fair-use principle, would seem to us to be *unfair* indeed.* Such practices could lead to dangerously reduced journal subscription runs, with unrecompensed secondary distribution carried on by these libraries assuming the principal role in eventual distribution to the users.†

This situation is likely to get worse as material is ingested into the massive computerized information retrieval systems now being planned, whose outputs are likely to be a complex collation of compounded elements from various inputs which fit the descriptor coordinates of the users' request. Identity of the original work for copyright recompense will be exacerbated. Naturally, the desire to employ new technology such as computer information systems, must not be repressed. The COSATI Task Force on copyrights (1967) has stressed this side of the problem, but unfortunately has not yet devoted adequate attention to the issues raised here in terms of support to the information dissemination process itself.

It is important to note that the authors of the journal materials are normally not concerned with financial remuneration;—as has been said earlier their remuneration comes from the recognition of the scientific community and the anticipated scientific feedback which results from their publication. Nevertheless, they are hurt indirectly by any action which

*Sophar (1967) has made a compelling review of these excesses. He points out that fair-use cannot be measured in terms of any individual use, but only in terms of the total use and total copying, which is seen to be enormous. "The conundrum of fair use is that the wider its area of application, the less fair and the less useful it becomes as a valid legal doctrine, and the more likely it becomes that the copier, researcher, educator and general user of copyrighted material, whether individual or institution, is an infringer."

†In a recent publication, prepared for the National Advisory Commission on Libraries, Clapp (1968) purports to show evidence that the subscription levels of journals has not been cut by the library reprography habits which he does not deny have enormously increased in recent years. But he completely ignores the basic issue which is the maintenance of the viability of these journals in an era of increasing numbers of worthy publications and an inflation of costs of publication. This work is also of interest in that it attempts to show that the copyright restrictions on copying were really not intended to cover the copying of material but rather to mean reprinting of the work for republishing!

compromises the economic viability of the journal literature since these factors reduce the resources available for journal publication and must circumscribe and constrain their flexibility in terms of editorial and printing support, speed of preparation and publishing, space available for full and clear exposition, number of pages per issue and frequency of publication (leading to backlogs of accepted works, etc.). In addition, the copyright protection that societies seek is also for the purpose of maintaining the integrity of the work with regard to its completeness and unaltered nature as well as its legibility. Copying often compromises these elements of the work. It should also be remembered that the income derived by professional societies from successful publications often helps in supporting other important scholarly and communication undertakings which would be difficult to support directly.

Of necessity, the journal publishers are increasingly forced to solicit other sources of revenue such as advertising, which tends to dilute the technical material. Furthermore, the financial difficulties of professional and learned society publications have already and could increasingly lead to the dominance in the scientific publishing field of commercial journals which tend to find other ways to maintain solvency, but whose reviewing practices and concern for the well-being of the scientific communication process are often secondary to the profitability of the operation. As has been observed, the power of a professional society to foster its own intellectual integrity, as well as to elicit the best efforts of its members in cooperating with all phases of the operation, is highly significant in keeping the operations intellectually honest and relevant. These self-correcting pressures are often absent in more commercially dominated enterprises.

It is worth looking at the costs of journal publication in some greater detail. At an ADI symposium, on the subject of Primary Journals, Mr. J. H. Kuney, of the American Chemical Society (1962), presented some data showing that of the major items in the cost of printing and distributing, approximately one-half is for type-setting* (which of course does not vary with the number of copies, as does presswork, paper, and mailing charges). At that time the cost of getting a single page to a single sub-

*OSIS has pioneered in supporting work to automate and lower the cost of these functions; e.g. cf. their work in the field of computer-operated typesetters.

scriber was one to one and a half cents, for journals with circulations between 6000 and 10,000 and $2\frac{1}{2}$ to $3\frac{1}{2}$ cents per page for subscriptions of only 2000. Thus, the break-even price for journals publishing 1000 pages/year is $10–$15 for the higher circulation and $25–$35 for lesser circulation. Applying an inflation increase of a factor of 2 to this, for 1969 prices, would easily explain the $50 and greater charge for many smaller subscription journals.

The Government has undertaken some support of the scientific journal literature. As discussed above, this includes approval of page charges for a limited portion of the preparation expenses of journals published by not-for-profit organizations. In addition, the Office of Science Information Service (OSIS), in NSF, has subsidized the gestation period and infancy of new journals where societies felt such undertakings to be worthwhile and potentially promising. Once organized, these undertakings have been expected by NSF to be self-sufficient; this is often achieved by wide use of advertising revenues, whose tax-free status has now been rescinded by the Internal Revenue Service.*

In view of the nature of the journal literature as a national resource, and by virtue of its threatened existence by unauthorized copying and the general escalation of preparation and printing and distribution charges, it is reasonable to ask whether there might not be a more equitable manner of supporting these activities which would provide a greater opportunity for expansion of needed services, viz. a larger subsidization. Of course, this must not be at the expense of objectivity or at the risk of domination by the government sponsor, since the democratic requirement of a "free press" extends just as strongly to the scientific literature as it does to other branches of journalism.†

Commercial publishers naturally have a quite independent attitude with

*In December 1967 the Internal Revenue Service announced, after hearings brought about at the probable instigation of commercial antagonists, that henceforth such revenues received by tax-exempt institutions (under section 510-c-3) would now be subject to tax if the subject journals were in areas where competition with commercial media was evident. It is a nice point whether this will affect purely scientific journals, in spite of the existence of commercial journals on similar subjects, since competition may be difficult to establish.

†It should also not be at the expense of the further proliferation of journal titles; there should be a strong expression of user needs and a reviewing process by scientific advisers before a new journal is supported.

regard to these matters. Supported primarily by advertising revenues* (which indirectly are provided by the overhead charges on government R and D contracts), they do not wish to see anything done to differentially strengthen the not-for-profit journals. Thus, page charges, which are currently limited only to the not-for-profit publications, strike them as unfair and, furthermore, they claim, carry an overtone of intrusion on the editor's prerogatives and the sanctity of the free press. On the other hand, they are in the forefront of those concerned with the reprography problem and the copyright issues.†

Clearly, a good deal of responsible thought must be given to the economics of information handling and the related utility issues raised above, in connection with the future of the three principal media; journals, books and technical reports. We recognize that to raise these issues is not to resolve them; however, because of their complexity, we can, at this time, say only that this is likely to be an international problem of growing significance which deserves immediate attention from the various community elements involved. It would appear that some innovations are needed in connection with the support of the communication media in light of the special problems discussed above. We are particularly impressed with the proposal of the Committee to Investigate Copyright Problems Affecting Communication in Science and Education (CICP) for a clearinghouse to expedite the effective and rapid utilization of copyrighted materials with full arrangements for fair recompense. It may also be possible through such an arrangement to utilize supplementary printing of additional copies of journal articles (reprints, for example) since the printing press still has many advantages of economy and legibility over the copying machines when demand can somehow be anticipated. The possibilities of selective dissemination of information through computer-oriented ordering based on descriptor or common citation coupling would appear to offer great potential in this connection.

*The significance of technical advertising in technical communication should not be overlooked. As Galbraith has pointed out in his recent book, *The New Industrial State* (1967), the industrial "technostructure" has a continuing need to stimulate demand and to interface with the "scientific and educational establishment". Advertising is their best tool, and to it they bring journalistic talent and resources in a manner which makes this technique especially effective.

†See, for example, the recent book *Nearer to the Dust* by Gipe (1967).

CHAPTER 4

The Technical Report Literature

As HAS been noted earlier, the scientific literature provides the research community with an essential connectivity and continuity with the works of others. Furthermore, it establishes the requisite discipline and rigor within the community through the mechanism of review and critique of their findings. For the technologically oriented community there is no comparable canonical literature, but the engineering community is clearly disciplined by the success or failure of their development or product. Its evaluation plays an analogous role to the literature in keeping the practitioner "honest". The other important parameter supplied by the literature is establishment of priority of effort—the technologist establishes priority through the market place and the patent system.

Nevertheless, there is a considerable body of what has amounted in the past to a form of "undergound" literature which has served the technologist in a comparable manner to the role of the journal literature for the scientific community. The vehicle that has been found particularly attractive for the conveyance of technical information in this new science–technology milieu is the so-called "technical report". It comes in many sizes, shapes, and forms, but its essential features involve the reporting of a coherent effort by the report's authors for the purpose of satisfying a contractual obligation of some sort. While we shall have occasion below to delve into many phases of the report literature, it is worth noting at this point that the principal exponents of the value of this literature have been the technologists for whom it appears to fill a vacuum and provide for a felt need, whereas the principal criticism has come from the basic research scientific community for whom it appears as a threat to existing traditional literature.

Weisman (1966) tells us that the term "report" derives from the Latin *reportare*, meaning "to bring back" and suggests a working definition as "organized, factual and objective information brought by a person who has experienced or accumulated it to a person or persons who need it, want it or are entitled to it".

Many have come to associate the burgeoning of the report literature with the federal government's support of research and development and the associated need to record the progress and results of the contractual obligation both during and following World War II (Miller, 1952; Herner, S. and M., 1959). While there is no question but that this has been an extremely important factor, it would appear that the format was and is widely used in industry to report results, independent of outside support. Furthermore, the limited circulation of a scientist's experimental and theoretical results in a form prior to journal publication, could well fit the category of technical report. (See the subsequent discussion of IEG's and preprint circulation.)

A behavioral scientist would stress the behavior–stimulus aspect of the report:

> More than any other form of written discourse, technical reports are prepared with the intent of providing the user with definitive information which will direct his immediate or near immediate behavior. More often than not, it tells him what to do and how to do it.
>
> The Technical Report, then, has this strong element of direction of user behavior. One might think of it as a condition of stimulus control of behavior (Ronco, 1965).

Government authorities would stress the highly detailed and complete nature of a report (compared to an abbreviated journal article, for example). Thus, the *DoD Glossary* (1964) gives this definition:

> A report concerning the results of a scientific investigation or a technical development, text or evaluation, presented in a form suitable for dissemination to the technological community. The technical report is usually more detailed than an article or paper appearing in a journal or presented at a meeting. It will normally contain sufficient data to enable the qualified reader to evaluate the investigative processes of the original research or development.

To the community responsible for generating, editing and handling the "canonical" journal literature, the technical report is frequently seen as a preliminary, unpublished version of a work, not yet passed through the rigorous review process, and unavailable to the scientific community in general; in short, not worthy of consideration as part of the "scientific archive".

In view of the wide spectrum of materials that fall into the technical report category, generalizations are particularly odious. In order to clarify the characteristics of the different species in this genus we have attempted a taxonomic classification, given below in Appendix A.

Regarding the statistics of the report literature, it is difficult to obtain any really significant figures. Comprehensive library archives contain almost a million such reports and the population appears to be growing at a rate of several hundred thousand a year. The Federal Clearinghouse for Scientific and Technical Information currently announces some 30,000 reports a year, but these are only the unlimited, unclassified reports that have been passed to them by other agencies. Including industrial contractor and subcontractor project reports, GPO technical documents and miscellany available from other sources, we estimate that some half a million items per year fall into this technical report category.

In view of the greatly different environment of the research and development worker, compared to the frontier basic scientist described above, we should expect to encounter different attitudes about the role and utility of the report format. Thus, research and development practitioners often see great merit in a number of attractive features of the technical report literature.* To many it "constitutes the first appearance in print of much of the newest and most important scientific information being developed today" (Gray, 1957). Of great significance to those employed in the "applied" portion of the "research spectrum", is the technical report's customary comprehensive treatment of an application, together with all of the ancillary information, that is so significant for anyone facing the problem of making *use* of similar techniques. This can often include discussion of so-called negative results; that is, theoretical or experimental avenues that were explored but found unproductive for one reason or another—a valuable time-saver for the investigator who wishes to profit from the experience of others. This should be contrasted with the stylized format of scientific journal articles, forced by the diverse subject

*See, for example, the DoD user surveys of government and industrial research and development personnel who give high ratings to the significance of this literature (Auerbach, 1965, and North American, 1966).

matter of their coverage and the rigors of economics to high density, abbreviated formats containing only the essentials of the effort in a manner which invariably conveys the impression that the investigator immediately seized upon the "correct" path of investigation through some "divine" inspiration.

Timeliness is another potential asset of the technical report, especially for those fortunate enough to be included with the select group on so-called "primary distribution" to receive the work. By virtue of the delays in announcement, indexing and secondary distribution, and the various limitations on this process, there are many for whom this factor is of no advantage, but for those members of the technological equivalents of academic "invisible colleges", information may reach them through the technical report route a year or more before wider broadcasting through an eventual journal article. The "Crawford Task Force" for the President's Science Advisor (1963) stressed this point:

> One reason for their (i.e., technical reports) continued existence (even) in unclassified areas is their currency. Information can be distributed to both performers and administrators more rapidly in this form than through more formal publication channels. There is an even more significant reason resulting from their value to development efforts; the technical report is a primary recording medium for applied R and D work. It is revealing that most of the criticism of this communication tool comes from the research scientists who are accustomed to the conventional scientific journals; the strongest defense for reports, on the other hand, comes from the technological community for which adequate alternative media are not available. . . . The technical report, whatever its shortcomings, is likely to remain an important element in the technological communication network.

As mentioned by Weinberg (PSAC, 1963):

> The documentation community has taken an equivocal attitude toward . . . reports. In some cases the existence of these reports is acknowledged and their content abstracted in the abstracting journals. In other cases . . . reports are given no status; they are alleged to be not worth retaining as part of the permanent record unless their contents finally appear in a standard . . . journal.

Libraries have never been comfortable with "technical reports" and have resisted their intrusion into their orderly premises. This attitude, while rationalized by their belief that all worthwhile material in the technical report literature will eventually find its way into the periodical literature, is probably more realistically due to the practical difficulties of cataloging and shelving this variegated species.

On the other hand, there are those documentalists for whom the technical report format is the archetype of the so-called "separate", which form has been advocated on its own merits for some time by information specialists. (Phelps, 1960, and Coblans, 1965). Thus, it can stand alone, be given a more or less "universal" reference number,* be separately stored and retrieved, and can expeditiously be made available in hard copy or microform in the local work environment of the user "at the time of need".†

Thus, we see that the technical report has been variously praised and maligned as a medium for scientific and technical communication.

The crux of the indictment, stated by those responsible for the canonical scientific literature, hinges on the lack of reliability of the report literature by virtue of its alleged unreviewed nature. Thus, Dr. Pasternak (1966), of the APS's *Physical Review*, feels that "the essence of scientific journal publication is the *orderly* communication of scientific information", which he apparently finds lacking in the report literature. He goes on to state:

> the refereeing contribution lies far less in the yes-or-no decision on publishability than in the service rendered by improving the papers ultimately published—through the elimination of flaws that would bother readers other than the referee (for example, misleading claims, omitted details, ambiguous statements, minor errors in argument, overlooked pertinent references, unrealized implicit

---

*However, this is a non-trivial problem, and much confusion has resulted from the multitudes of series of reports, so that the Special Library Association (1962) has had to prepare a dictionary or Coden (analogous to that for journals), which see. It is hoped that the trend toward central clearinghouses, and required standard format improvement by COSATI's operational committee (1967), can solve this problem. Dr. Sherwin (1968) has proposed a national (and then an international) standard for reporting current efforts and their reports, which would clarify this matter (viz. a Governmentwide Form "1498" system).

†This is especially significant in view of the delays, so often encountered in retrieving material in journal or monograph form, which are customarily borrowed in bound volumes and often not available for other users, who are then given the very unsatisfactory reply: not on shelf, on inter-library loan, or the ever-frustrating "out to the bindery", in place of the information they seek.

For a discussion of the attitudes and procedures of the library community that handles reports, cf. SLA's Regional Workshop Proceedings (1967).

assumptions, obscurity and discursiveness). The result is an article that is easier and quicker to read in detail and to understand and use (or even to decide not to use) and one that is more reliable than the original preprint. It is this filtering procedure that makes for the orderliness of communication through scientific journal publication.

Thus we observe that to many in the community of professional scientific journal editors, the report literature represents a threatening wave of "unreviewed" material, of doubtful quality, which fattens itself at the trough of government subsidy whilst their "high-priced quality spread" suffers increasing financial difficulty. In short, they fear that the massiveness and widespread growth of this "debased" technical report literature will, by some analog of Gresham's law in economics, gradually drive the scientific journals out of currency. Their explicit policy in dealing with technical reports is somewhat schizoid—they often refuse papers submitted to journals if the contents have previously appeared in technical report form, on the grounds that the work has already been "published"; while, on the other hand, they often refuse to permit papers in their journals to reference technical reports on the grounds that they are not generally available and therefore represent "unpublished" material for which the grossly unsatisfactory reference "private communication" has been adopted.*

Criticism of reports has not only been confined to journal editors. Thus, the "Elliott" House Select Committee on Research (1964) called attention to this comment by Dr. T. J. Kennedy, of NIH:

> These documents should not be assigned a value beyond their real worth. They certainly should not be permitted to become part of the archives of science, a destiny that is de facto guaranteed if the agency publishes them in any form whatsoever. Nor should they be permitted to increase the already large

---

*While there is a tradition (backed up by some international support from the ICSU Abstracting Board) that referenced material should be available to all, there are two counterpoints on the side of the argument for referencing of proprietary items.

1. Even though a report is limited, some people can have access to it and why discriminate against them (especially important where national security limitations are concerned).

2. Priority should be established—it is true of course that this could be done by reference to personal communication, however, this is somewhat less conclusive than a publication, even one limited in availability.

> volume of published literature, with which the scientist must cope, and which the documentalist must collect, index, abstract, store in retrievable form, announce, and distribute.

Government advisory groups have acknowledged that the

> quality of a report can only be ensured through an adequate review process. In part, this review is the responsibility of the supervisory structure. Very vital additional review is provided by criticism from the scientific community. If the latter is to prove effective, the reports must have adequate dissemination (Crawford, 1962).

Unfortunately, this traditional principle of review and criticism comes into conflict with the requirements of various sponsoring groups to control the dissemination of "their" reports for proprietary or security reasons (real of imaginary). It is one of the cardinal facts of modern scientific and technical life that scientific and technological information has become so significant to the business of individuals and institutions, the economics of industries and nations, and the national security of peoples, thus requiring that it be "managed" consistent with the awesome consequences of its availability. It is one of its distinguishing assets that the technical report has proven to be a very effective medium for the packaging of "sensitive" information by virtue of its manageability. It permits the circulation of information to an authorized, controlled group according to what has come to be somewhat euphemistically known as the "need-to-know" principle. Proprietary information has been handled this way in "company confidential" reports for many years and security classification of information and its inhibited handling has become a way of life for a large fraction of the world's scientific population. Accepting the realistic constraints that proprietary and security requirements place on the communication process, there is still much at issue in the intelligent implementation of these requirements. The boundaries for the requisite controls are difficult to draw and the issues are complex indeed and are likely to stay that way. Even within the closed communities that exist with access to proprietary and security classified materials, there is still a strong need for scientific criticism and self-correction to occur, if scientifically valid programs are to take place.*

*Cf. the cogent criticism, with some telling examples, by Dr. B. Commoner who contends that this is not possible unless much wider discussion occurs (1966).

In view of the criticism that has been levied against the technical report literature on a number of these counts involving quality, it is considered worthwhile to discuss the various issues and policies in some depth.

As has been indicated above, the technical report describes the results of a contractual effort which a research and development team has performed. Their qualifications to *perform* the effort have been previously scrutinized and screened by selection committees. Often this selection is based upon a proposal evaluation, past performance check, discussion with previously related government project monitors, etc.* As the work progresses there is usually considerable coordination with other groups active in the field, circulation of progress reports among related project officers and review by government management and advisory groups.

Upon the successful conclusion of the research (often projects are extended and funded for additional work until *some* successful results are forthcoming), the final technical report is prepared. This report, the product of a number of co-workers at the industrial or other laboratory institution, is then reviewed and edited within the parent institution. Because of the important institutional image factor and the financial significance of the contract relationship itself, considerable effort is usually expended in the report preparation (time, drafting and photographic assistance, technical editorial support, etc.), and management quality control is exerted to make the finished draft report a quality product.

Many government groups require that the draft report be submitted to them for further review prior to correction and final printing. This stage often includes preprint circulation for critical comments by the sponsor among his colleagues and advisors.

When the final report is corrected and printed, it is then circulated to the government sponsor's primary distribution list, as well as to central distribution points for announcement, indexing, abstracting and wider dissemination. Specialized information centers often make a separate announcement of the report's availability, often with additional critical annotation. On an increasing but selective basis, pertinent technical reports are also being abstracted and announced through the scientific secondary journals (*Biological Abstracts*, *Chemical Abstracts*, etc.); as cited

*It is interesting to note that the Government is rarely criticized on technical grounds for this screening and selection process of a contractor, as well as the subsequent direction of the effort and expenditure of funds in the R and D process itself.

by the journal literature they are also included in the *Science Citation Index*.*

Thus we see that there is no prima facie case that can be made in general criticism of the technical report literature's quality. What has been described above is admittedly the system functioning at its best; there are, however, several areas where pitfalls can occur. In view of the fact that a great deal of the R and D effort has as its goal the supplying of hardware or the development of a technique, it is usually the *performance* that most interests both the contractor and the government project officer. The report itself comes after this performance in priority as well as in time.† Thus the question of adequate and high-quality documentation, as well as the proper review by all in the management chain, are accomplished with dwindling time, interest and resources as the contract is completed. As a practical matter, the contractor is probably applying considerably more time and effort during this stage with proposal writing for the follow-on effort or has reassigned and dispersed the original team to new funded efforts.‡ Furthermore, as has been indicated in an earlier discussion of the facts of technological life, the engineering team is not particularly anxious to disseminate to its potential competitors the details of its accomplishments which it still may consider proprietary. The project officer is by definition project-oriented—he is striving to accomplish his assigned tasks, and acting as an equivalent Journal Editor is not usually considered to be one of them.

Thus we see that the environmental conditions for reports do not at all correspond to those in the case of the canonical literature, and it would not

*The symmetrical tool for covering the citations of the Technical Report literature has not yet been provided; we think that it has much to recommend it. Although there has been some a priori feeling expressed that this literature does not exercise the same respect for prior work, the facts of the matter are apparently that the report literature averages a larger number of references per document than journal papers do (seventeen as compared to fifteen, per private communication from Dr. E. Garfield).

†It is significant that those organizations most concerned with the quality of their reports are those for whom the report represents the principal product—such as the basic research and systems analysis type organizations.

‡It is interesting to note an unfortunate analogy with the modern university research team; too detailed a request for a paper's revision by a journal editor may meet with the author's plea that his research assistants have graduated and the paper will have to stand as submitted!

be surprising to find major deviations in quality. However, it is difficult to compare such a subjective parameter as quality between any large groups of the technical literature. A number of opportunities exist where the two bodies of technical reports and scientific journal articles intermix, which can serve as a base for comparison. Many technical reports are submitted to journal editors for publication as papers. There is some evidence* which shows that the subsequent journal paper, delayed by a year or so beyond the time of primary distribution of the original report, differs very little from the report. If anything, the report, because of its enlarged discussion of the work, would often prove a more valuable reference to the user.† Although many journal editors have a policy of rejecting material submitted in report form (for a number of reasons, one of which was undoubtedly that the criterion of original publication was not met since the report was previously "published"), there is also evidence that with the very large and redundant journal population in existence, a persistent author can eventually be "published".

The subject of technical writing, as a problem area affecting the quality of the technical literature, has frequently been raised. Weinberg (1963, 1965) has called repeatedly for greatly increased attention to this problem. This would appear to us to be an educational problem which cannot yield merely to exhortative criticism, but rather requires basic curriculum improvements. Recently, Prof. Woodford (1967) has stressed the importance of "Sounder Thinking Through Clearer Writing" and has pointed out that this subject is not just of academic interest but actually affects the soundness of the scientific method as well.‡ John Maddox, the Editor of *Nature*, has also pointed to the possible improvements in the readability and assimilation potential from improved styles in journal writing. These could really help decide "whether the scientific literature is dead or alive", as he appropriately entitles his article (1967). "Why", he asks, "cannot

*Cf. Garvey and Griffith (1965).

†One could say, somewhat facetiously, that what the report may lack in quality it makes up in completeness!

‡Many have pointed out that this is more easily said than done. In fact R. H. Good (1967) points out from his own frustration in being forced away from simple and direct prose into more elaborate impressive but empty phrases, "that scientific writing is the way it is because its readers actually prefer it that way. People's actions do not always correspond to their words. Everyone is against sin and bad writing, unless given a free choice."

there be some experiments in which reports of scientific work are written deliberately with the objective of bringing enlightenment, pleasure and occasionally excitement to as many readers as circumstances will permit?"

As has been mentioned, the growth of specialization is a well-recognized phenomenon which is probably an essential part of the solution of complex decision making problems. Yet the effects of specialization are to create specialized terminology and jargon, to limit the area of direct communication, and to make accessibility outside the specialized field increasingly difficult. Although data in one area may well contain information relevant to a decision in some very different area, the specialization of both the subject field and the language raises almost insurmountable barriers against this relevancy. There are several tools available for coping with this "tower of Babel" syndrome. The advent of good literature terminology control through the availability of an up-to-date Thesaurus, or subject terminology authority (cf., for example, the recent DoD Project LEX Thesaurus, 1967), can assist in communication across disciplines and in coupling between specialists.* Also, coupling between specialists can be expedited through tools such as the *Science Citation Index* which eliminates the descriptor indexing technique which is dependent on the adequacy of the Thesaurus and the intellectual dexterity of the indexer. Of course the use of tutorial reviews to assist in this regard is most significant and will be expanded on further below.

It would be difficult to generalize in ranking the journal literature versus the technical report literature from the standpoint of technical writing or style. While the journal literature is subject to considerable review and editorial suggestions, the institutional nature of the technical report offers the opportunities for much greater support by professional writers, editors and draftsmen so that the work itself can achieve a much greater proficiency. Manuals and books have been written to assist in the preparation

*"Each specialized area of technology usually develops its own specialized vocabulary or jargon to express the concepts peculiar to its own interest. Analysis of these specialized vocabularies reveals that many or most of the terms thought to be unique to a particular speciality are essentially synonomous with, or clearly related to, terms used in other areas of technology. Moreover, it is quite common for technologists in one area to borrow terms from another and use them for their own purposes. The Thesaurus shows these relationships and thereby permits the selection of terms that will improve communication both within and across boundaries of engineering and scientific specialities" (DoD, Thesaurus, 1967).

of technical reports (Weil, 1954) and (Weisman, 1966),* whereas the literature on journal article preparation is strangely scarce. Actually, the most "professional" technical publications are usually those of the trade journals, where the articles are prepared in the first instance by technical journalists and exhibit an attractive and clear style which is undoubtedly responsible for their popular appeal. Although we would all share Dr. A. Weinberg's hope that good writing habits and training should become more widespread, we believe that this element is not uniformly found and we believe that commercial journals will continue to have a preponderence of this scarce commodity.

The problem of quality control for the "classified" literature presents special problems. While the need-to-know limitations must, of necessity, constrain the audience, the significance of this work demands even greater attention and concern for good quality control practices. This requires the widest availability of this material to those within the cleared community for whom the material may be relevant. This presents a ıeal challenge to the Government; we believe that considerably more attention must be given to it in terms of the amenities of good communication practices, but this is not the place to go into all facets of this difficult problem in detail.† We do feel that concern for "secrecy and dollars",‡ as well as the awesome consequences of actions taken by the Government under conditions of secrecy, deserve the best and continued efforts of our community.

It should be pointed out that the fact that technical information through the technical report channels is controlled by the corporate institution (reports, unlike scientific articles, are products of an institution, often the authors are not mentioned at all or credited only in a foreword) in the first instance, and then by the sponsor in terms of modification, release and dissemination, poses a serious potential danger to the integrity of the communication process. All too often these controlling groups act in an analogous fashion to the concept of a "Maxwellian Demon" in statistical

*This has even included attention to the psychological factors affecting report reader behaviour; see, for example, Ronco (1965).

†Almost all studies of government communication problems have recommended greater attention to this problem, but the clearing of these Augean stables awaits a Hercules.

‡See the recent article with this title by Simpson and Murdock (1967).

mechanics, who can, through the positive and negative release and dissemination policy, control the image of a program or policy to satisfy their own inclinations. It is this potential control that poses the greatest danger to further efforts of the Government to offer subsidization to the scientific literature; clearly a better and safer way must be found to provide for the economic vitality of this national resource. Any technique by which the support is "decoupled" from the subjective constraints of any particular technically oriented sponsor is to be preferred. It is for this reason that support through NSF's OSIS is very much preferable than from one of the development agencies; however, the facts of economic life seem to associate substantial funds only with specific R and D projects.

The Government, as the sponsor of a tremendous amount of scientific and industrial research, is primarily interested in the exploitation of this information for the specific problem at hand in its explicit contract goal. As a philosophical principle, it now recognizes the potential wider significance of the particular effort to other groups and other problems. Thus the making available of technical reports issued under a contract obligation through clearinghouses, etc., is clearly a positive step in the dissemination of information which might not be expected to voluntarily occur through the "open literature" in all cases without this intercession. Thus, for example, Dr. Shockley expressed his opinion, in testimony before Congress (Senator Hennings' Judiciary Committee, 1958–9), that scientific information dissemination in the solid state physics area is best where the Government support of projects enforces distribution of results, as compared with those private areas which remain company proprietary property. There is also some indication that governments in other countries wish to emulate the United States in expanding their technical report literature as a way of promoting greater industrial cross-fertilization as an aid in overcoming the limitations of the so-called technology gap. (See, for example, the admiration that Lord Mountbatten has expressed for the potential of technical reporting in the U.K. electronics area.)

A difficulty arises, however, in that the compartmentalization of research support through numerous project offices and agencies results in widely different dissemination policy interpretations. Where agencies have statutory responsibility for information dissemination, or an agency administration aggressively pushing in this direction, the situation is far superior to those cases where such policy is generally ignored or only given lip service.

The subject of the transition of works from the report medium to that of the open scientific "canonical" literature is a complex one. The initiative is, of course, with the individual authors. They and their institutions have a definite incentive to move in this direction from the standpoint of wider publicity and renown for their work. Feedback from "peer groups", while not nearly as prevalent in the literature as the traditional mythology would imply, is nevertheless considerably greater from publications appearing in journal literature than from the report literature. The significance of requests for "reprints" in the journal literature was a technique for identifying individuals with *a priori* common interests (now overtaken by institutional copying). Report literature, on the other hand, is usually secondarily disseminated through central depositories.

> Because the secondary distribution of journal articles from sources other than authors, who are probably the major source, is highly decentralized, no figures are available to compare with report clearinghouses. Reliable data on the number of requests received by authors for reprints of journal articles are scarce, but one recent study established the median as between 11 and 15 (Fry and Associates, 1962).

While much technical report material does appear subsequently in conventional primary journals, quantitative data on this point are meager.* A pioneering study on this subject, made in 1955 by Gray and Rosenborg, found that:

> For the approximately 1,100 reports whose case histories were studied: (i) that 60 to 65% of these documents contained publishable information, as judged by their authors; (ii) that for about half of the reports that contained publishable data, all such information was published, but some of it, with a 2-to-3 year time lag; and (iii) that the publishable data in about one-fifth of the reports probably never appeared in conventionally printed form.

In a 1960 study, Herner found that:

> of the 2,295 specimen reports 495 or about 22% had been published in whole or in part. This figure bears a striking resemblance to the figure of 21% obtained by Gray and Rosenborg in their earlier study (although they limited their study to

*A useful survey of the policies of government agencies on this question, as well as their data on the subject, can be found in the House Select Committee on Government Research (Elliott Committee) Report (1964).

DoD reports). The 495 specimen reports that were published in the present study were found in 161 different journals, 32 books, 6 patents and 47 meeting preprints. Forty-two of the specimen reports (1.8%) were actually reprints of papers that had already been published.

The bulk of the specimen reports from Government offices that were published were published in 7 to 9 months from their issuance. Reports from the other three sources were published in 10 to 12 months on the average (viz., colleges, industrial firms, and independent research institutions). . . . In regard to type of research reported, as might be expected, in view of the fact that publication is its main physical product, reports resulting from pure research were published with far greater frequency than reports on applied research or development (2:1 in favor of pure research).

The most recent study of this transition from the report to the journal literature form has been the impressive communication study within the field of psychology by Garvey and Griffiths of the American Psychological Association (1965). It's results are consonant with those above, and stress the significant time delay between the two media.

The Government, in explicit recognition that a research effort is not completed until the findings are published, has authorized page charges for journal publications from contract funds. This was authorized in a 1961 action by the Federal Council for Science and Technology at the initiative of Dr. Adkinson, head of OSIS of NSF. The payments were predicated on the understanding that such charges would be levied on all contributors equally and that they should, of course, not be a prerequisite to publication. A further stipulation that the journal be a non-profit publication seems to be on somewhat less firm ground as is a restriction of the payment to defray the "intensive" (i.e. independent of distribution) costs of journal publication.

But the acceptability of manuscripts to journals, where the material has previously appeared in technical report format, is quite variable, as has been noted. When primary distribution has been limited to less than a hundred copies of a mimeographed or "processed" report, editors have generally agreed to ignore it,* and to accept the manuscript for consideration. With the modern "widespread" dissemination of report literature, no such "unpublished" attitude can be taken, and it is understandable if editors of journals with large backlogs of original papers waiting to be

*Agreed to at a 1950 meeting of the National Academy of Sciences attended by representatives of government agencies and the editors of a number of journals.

published, and severe economic constraints, were to give less priority to manuscripts based upon reports already widely disseminated. However, the serious question arises as to whether the report format constituted publication from the standpoint of quality review and issues of priority. We are thus left with something of a circular problem.

When it comes to legal questions involving priority of efforts appearing in technical report form, there are also some uncertainties. Those reports that are available from central clearinghouses such as GPO or CFSATI are clearly in the public domain for patent and other legal purposes. Reports otherwise circulated are not so considered and their status is moot.

It is interesting to note that the present efforts to liberalize the availability of government contract reports may be adversely affected by the proposed revisions in the Patent Law (President's Committee on Patent Policy, 1967). Under the revised policy, as part of the "first to file" approach to patent awards, the one-year grace period, during which scientific publication could precede a patent application under the old system, will be rescinded. Thus, disclosure of scientific and technological facts in a technical report made available for open sale through the Superintendent of Documents or Commerce's Federal Clearinghouse for Scientific andTechnical Information will, in the future, abrogate any proprietary rights for subsequent patents. We anticipate that this will lead to new limitations on technical report distribution to preclude this from happening. On the other hand, the new policy also stipulates that the patent disclosure is to be publicly disseminated after 2 years, thus avoiding the present even further delay until patents are actually granted.

With regard to the economics of government-sponsored reports, the situation is currently in a state of ferment. Since this literature arises primarily in satisfaction of government contractual relationships there is no question that the Government has a responsibility for proprietorship over its distribution. For the use of government related efforts, there have been established agency clearinghouses for secondary distribution and abstracting; indexing and retrieval services (viz. AEC's Technical Information Division at Oak Ridge, NASA's Information System, operated under contract by Documentation, Inc., and the Defense Department's DDC (Defense Documentation Center)). Other government agencies and contractor personnel have usually been eligible for "free" servicing of their report requests through these agency clearinghouses. In the current shift towards the encouragement of microfiche many of these centers are applying pressures in this direction by restricting free distribution to this format.*

With regard to further distribution to the "public" (i.e. the non-contract related groups) the principal avenues for dissemination are through the Government Printing Office and the Federal Clearinghouse for Scientific and Technical Information (GPO and FCSTI, respectively). While there is not always a clear-cut distinction between these two points of dissemination, the techniques for handling are somewhat different. Thus the GPO sells to the public from its supply of overprinting for the original "official" distribution of the material. As a result, their charges are quite nominal. The FCSTI responds to public requests by means of individual, on-demand reproduction from its microform masters once its small stock of "extra" copies received from the cognizant government sponsor is exhausted.

The question of copyright in connection with government-sponsored works is currently being widely discussed. It is hoped that this situation will soon settle out (perhaps by means of the accomplishments of the recently proposed National Commission on Copyright Issues), since the current situation leads to confusion and inhibition on the part of the community. It has been the policy to exempt government works from copyright. However, this has been narrowly interpreted to concern only government documents or the official works of government employees.† The writings of contractor personnel has been copyrightable, although the reports which are submitted in fulfilment of the government contracts are used by the Government freely to satisfy its own and the public's needs. Many agencies have recognized the superior distribution and marketing mechanism offered by commercial publishers, and have found it desirable to have the final report product published under copyright by the publisher and available to the public through their marketing facilities with the government distribution being made up by copies supplied to

*Note, for example, the recent movement towards user charges by the DoD (*Defense Industry Bulletin*, 1968).

†The recent celebrated Rickover case, which was taken to the U.S. Supreme Court, has led to the interpretation allowing copyright for the work of government personnel done in their own time or not otherwise connected with their official assignments.

the Government as a royalty substitute. Where the manuscript was fully sponsored by the Government, it has also been policy for the government administrator to hold the copyright but to arrange publication through a private publisher.

A matter of recent significance, related to reprography, etc., has arisen in connection with the common practice of scientific journal publishers to copyright materials, most of which it must be admitted, have resulted from government-sponsored research efforts. Some agencies have insisted that the publication include a statement recognizing this prior contractual right of the government sponsor to the resultant work, for copying for any government use. Journals have naturally resisted such *carte blanche* statements and have rejected papers where this was given as a prerequisite to publication. Some government agencies (e.g. FCSTI) have attempted to sell journal reprints of government-sponsored works without authority of the journal.

While there does not seem to be any question that the government sponsor has purchased the research results reported upon, the fact that the journal has added to the conceptual phrasing of the original technical report and has reviewed, edited, published and indexed the work should not go unrecognized. We believe that the Government should, therefore, take no steps to financially injure the journals but should rather attempt to satisfy its needs for secondary distribution through the journal standard access procedures. This will eventually serve the Government's needs best by enhancing the entire community's improved access to the quality literature.*

In considering the somewhat competing relationships between the journal literature and the technical report literature, a number of suggestions have been made for improvements. We have found it heuristically instructive to consider two alternative hypothetical *extremal* approaches to the present dilemma of the multiplicity of competing media and the confusion that this introduces into the orderly communication system:

1. This "solution" would explicitly acknowledge the professional scientific journal literature as the preferred medium for the transmission of all types of scientific and technical information. It would reduce the

*See the related discussion earlier on the central library reprography problems.

technical report literature to the bare minimum necessary to *manage* the government research and development programs and would require the government sponsors to apply all reasonable pressures and incentives toward the prompt submission of substantive research results to the journals for review and publication where warranted.

2. This "solution" would acknowledge the technical report literature as the principal and preferred medium for detailed technical communication.

Under the first approach the scientific journals would require a large subsidization to expedite the handling of additional information which would be moved in their direction for primary dissemination. A much greater effort on the part of societies to provide the complete secondary bibliographic handling of this material, would also be required. New journals, both to include new specialized fields and to accommodate the technological type of information, would be required. Of course, some compromise might be possible by employing the journal review process to accept a work for publication by printing a precis of the article while merely referencing the bulk of the material (preferably, edited and reviewed, as well) which would be placed on deposit for demand retrieval. This would be quite similar to the auxiliary publication system first proposed by the late Mr. Watson Davis of Science Service (and organized between ADI and The Library of Congress), but presumably on a much expanded scale.

The principal difficulty that we see in this first approach is that of persuading the government agencies, who are responsible for carrying out research and development, that this system can adequately meet their needs.* They would become quite dependent upon the traditional communication channels for the detailed discussion of their own programs, and would undoubtedly need to reconstitute an abbreviated reporting system which might again grow into a new report literature. Of course, this dependence would give government a new incentive for being assured that the scientific and technical communication system was healthy and adequately supported to handle the publication requirements in an effective and timely fashion. It would also prove, in a more direct way than at

*It is true that several "granting" agencies (as opposed to mission-oriented agencies), already have adopted this approach.

present, the relevance and importance of the professional society involvements of individuals in government, industry and universities, whose time and technical devotion to these responsibilities do not now necessarily receive adequate approbation consistent with the importance of their work. Certainly the classified literature would have to be treated in a separate manner; however, here again the same principles of publishing could be used with the proper need-to-know constraints. A precedent already exists in this direction with the existence of security classified journals in a number of technical areas. (See, for example, *Proceedings of IRIS* (*Infrared*), *Journal of Ballistic Missile Defense Research*, etc.)*

The second approach is one in which we would clearly see a much greater and direct involvement of the Government with the literature publishing and dissemination functions. This seems to be the present natural evolution of agencies who take their information stewardship seriously and who aggressively attempt to provide the necessary primary, secondary, review materials and information retrieval systems for the research and development tasks of their own personnel as well as those of their contractors. This appears to be especially true when resources are adequate for major data handling facilities to be brought to bear on the problem. This can and does extend to the informal communication of results and observations through symposia, and in the rapid publication of proceedings,† all under the cognizance of the sponsoring government organization. The appeal of this approach appears to lie in the possibility for improved financial support because of the direct coupling with sponsoring and fund disbursing organizations as well as in the need for coping with the rising growth of "privileged" or "constrained access" information, which seems to be much more than a temporary phenomenon.

We would hope that under this latter approach the important characteristics of review, good and precise scientific writing, connectivity with

*Dr. Garwin (1968) has recently advocated increased emphasis on the journal format for classified and other "closed" government proprietary areas, for the principal reason of achieving objective, cumulative reasoning on the important issues and programs. This provision for continuity is especially significant to allow R and D to build on what has been done in spite of all the turbulence in personnel changes.

†These functions need not necessarily be performed "in-house" but may be contracted for; the important distinguishing area is the direct planning, supervision and control of the various functions.

related research through citations, and the other amenities of good scientific communications, would not be neglected. Nevertheless, we would still anticipate that many individuals in industry and government would prefer to utilize completely open channels for publication; as we have seen, such publication clearly results in greater community recognition and encourages technical discussion and feedback criticism in a way which seems beyond the capabilities of the report literature.

A basic objection to this second approach stems from the important concern that the by-passing and reduced influence of the open-literature professional-journals can lead to serious dangers for the integrity of science as well as the wise use of science and technology.* In an analogy to the public press, the free and open scientific literature plays a vital role in the self-correcting evolution of the scientific process. The informed technical public cannot abrogate the responsibility for analysis of the details and consequences of science issues and policy to any small closed circle of men, however brilliant and enlightened, without the proper open opportunity for wider review, especially in these days of interdisciplinary complexity, knowing the awesome and irreversible power of the tremendous forces available through science and engineering.

We are, therefore, convinced that neither extreme solution is tenable, practical or desirable. The proper satisfaction of the various technical requirements that exist will, we believe, necessitate a continued coexistence of the various scientific and technical communication media. We believe, however, that the particular strengths of each medium should be explicitly recognized and nurtured. The remaining overlaps and redundancy should provide a useful opportunity to permit innovation and improvement, and all competition is certainly not to be discouraged. For the scientific journals, we believe that their purpose remains the quality, orderly and open publishing of scientific advances for professional societies and related audiences, and the scholarly and archival recording of the growth of knowledge. We anticipate that such journals will continue to persevere through the contributions and volunteer support of their membership. However, since they clearly represent a national resource, we believe that the Government must endeavor to ensure their viability

*For a cogent discussion of the significance of open publication and the integrity of science cf. Commoner (1965, 1966).

and health through generous support on a non-interference basis. National Science Foundation support, more generous page charges, clear recognition that society service is to be considered worthy effort for industry and government approbation, permission to enjoy tax-free status for technical advertising revenues, copyright justice, etc., should all be strongly encouraged. This literature is clearly an important element for the success and well-being of science and technology in the United States and should be treated consistent with the valuable resource that it represents. While we have pointed out the greater *a priori* qualities of objectivity and excellence of the professional society literature, we cannot generalize in stating its universal superiority so as to completely justify favoring its treatment relative to that of the commercially supported journals. Where commercial services have established quality materials, such as recognized journals and review series, we believe that they should be treated comparable to society products, with analogous procedures for encouraging their vitality. We believe that each case of support will have to be judged on its merits. Government and industrial administrators should recognize the significance of the scientific literature by offering the requisite incentives and otherwise promoting and encouraging its use for the dissemination of government-sponsored scientific and technical information to wider audiences and for archival purposes. Thus, wherever possible, the Government should encourage submission of relevant scientific findings to scientific journals by approving the time and effort expended to do so under the terms of the project support. Furthermore, the journals, books and technical information services relating to the subsequent handling and retrieval of these materials, should be "allowed expenses" under contracted R and D efforts.

For the technical report literature, we believe that the Government should insist upon full and high-quality reporting of work done under government contract. Adequate time and resources under the contract must be allocated to the review function. Concern for the significance of the results of an effort and their dissemination to a greater audience than those close to the immediate project, demands that this review include a concern for the writing, editing, data, findings, dependent references and objective statements of the context and limitations of the results. Dissemination and concern for evaluation and review should be explicitly added to the responsibilities of government project officers as an assignment from agency directors and administrators for which they will be evaluated in addition to the specific performance of their project effort.

Under these improved circumstances, we believe that journal editors will feel a greater confidence in authorizing reference to the report literature for expanded treatment or specialized details and data or information that can only be made available on a need-to-know basis.

CHAPTER 5

Informal Information Exchange

As HAS been mentioned earlier, there is a much greater reliance upon informal media for diffusion of scientific knowledge at the frontier level than is obvious to the uninitiated. It is well known to institutional managements that, while new knowledge is made available to scientific and technological communities in many ways, the "exciting" new information becomes known by the *peer groups* in the world scientific community much sooner than it is known by other groups. Thus, important new knowledge is known to members of the "invisible colleges" well in advance of any formal written communication. (A classic example of this phenomenon is the case of the U.S.–U.K. awareness of the feasibility of uranium fission (discovered by Hahn and Strassmann in Hitler's Germany) through the scientific grapevine (Bohr *et al.*).) For a fascinating account of the invisible college at work in "molecular biology" see the recent book *The Double Helix* by Dr. J. Watson (1968).

This necessity to belong to the peer groups for early entree to the scientific frontier is a great justification for applied research groups to retain resident basic research staffs on the premises, to serve as "ambassadors" to the scientific information world.

The desire for informal communication of nascent scientific findings and preliminary comments, which scientists have, leads to the provision for a number of channels for informal communication. Thus, opportunities for expression and feedback exist in the interactions which an individual has with his local colleagues, in visits to other groups and at scientific meetings, and in correspondence with the more distant members of one's "invisible college". (See the earlier description of this concept p. 12.)

However, becoming aware of the individuals who are working along similar research lines to one's own is not an easy undertaking, especially in view of the recent proliferation of R and D activities in the country and in the world. This is an especially difficult problem for individuals shifting specialities or undertaking a new line of inquiry which takes them somewhat beyond their speciality. There are a number of innovations to achieve greater visibility over this "current effort" phase of information handling.* For basic science, the concept of a "Science Information Exchange" has been developed, very appropriately under the aegis of the Smithsonian Institution. Under the direction of the Federal Council for Science and Technology a governmentwide system for registering work units is being developed.† In both systems the emphasis is to employ some deep-indexing system and ancillary mechanization to assist in the retrieval of investigator information pertaining to an inquiry relating to his field of endeavor. Once this information is found, the seeker is encouraged to make direct contact with the other related efforts to coordinate and communicate and hopefully to avoid unwitting duplication of research.

This importance of personal confrontation for information exchange is not always well recognized in attempts to build up complex information systems. Nevertheless, user surveys have "found consistent biases by engineers towards the use of personalized, informal, oral means of technical communication—as opposed to more depersonalized, formal and written channels. . . . People are still the most effective systems for transmitting technical information to other people" (MIT, 1968).

Other elaborate information exercises (e.g. Air Force Project Forecast) have observed that

> there is no substitute for making intellectual contact with knowledgeable people for specific answers, referrals and data. With all due respect to formal bibliographic literature control and mechanized data retrieval there is just no substitute

---

*It is interesting to note that Congress has been one of the important groups pressuring for better handling of this information. Thus, for example, the "Humphrey committee" (1962) stressed the significance of this information as basic to good management practices (cf. the "Wenk" and "Stern" reports commissioned under their cognizance).

†See the work of Dr. Chalmers Sherwin (1966) for OST. Using the so-called form 1498 as an example, a government-wide current effort standard has been promulgated and a system is taking shape. Similar systems are operated by the State Department-External Research Division relating to social and political science research efforts.

> for a half-hour conversation with the expert in the field! This is so because the individual possesses the mechanism to integrate past experience, filter it, respond selectively to the question, and can be interrogated for further elucidation. There are also more subtle reasons stemming from the informality or oral communication as compared to the conservative stilted language of scientific documentation. One can learn more from the smile on a briefer's face when asked a pointed question about the reliability of his equipment, than one can glean from a 500-page glossy proposal (Passman 1963).

This significance of personal communication is a recognition that technical information is really a "vector" quantity involving direction as well as magnitude. The direction aspect involves status of a finding in terms of qualifications and future directions of pursuit of a field which are not often stated in the stilted and formal medium of the journal literature. Furthermore, as has been pointed out by Woodford (1967), scientific publication is subject to many delays, not the least of which arises from the feeling by scientists that preparation of manuscripts is dull and uninteresting once the excitement of making new discoveries has subsided. Publication will never entirely eliminate duplication of effort or provide the stimulus of personal conversation.* Personal conversations between scientists have always been a vital link in the development of scientific thought. Not only do active minds stimulate each other but there are details of experiments of theories, and of possible interrelationships which go unnoticed unless scientists can see and talk with each other personally.

Scientific meetings have become especially widespread in our affluent and gregarious society. There is now some concern that perhaps the self-regulating mechanism of the scientific community has gotten out of adjustment, as witness the great proliferation of such meetings. The economic problems of balance of payments has recently caused the U.S. Government to take a much more stringent attitude with regard to the support of travel to international meetings, and general budgetary pressures have also brought about greater scrutiny of attendance at U.S. meetings as well.

We all recognize that the formal papers delivered at meetings are only

*As has been noted by Garvey *et al.* (1965): "The most striking feature of the process of dissemination in psychology is how small a portion is easily available to the scientific community. The public dissemination of information occurs late, takes only a few forms, and, in its complete archival presentation, that is, in scientific journals, has a small immediate audience."

one phase; some say not even the most significant phase of the information exchange process conducted at meetings. While there is often an expression of fiduciary concern to publish conference "Proceedings",* the face-to-face contact of individuals and the opportunities for informal discussions which these occasions provide are most significant for the exchange of really significant information on new developments and directions for future research.† With regard to these face-to-face opportunities for direct interaction through informal communications, Garvey and Griffiths (1967) have pointed out that:

> the interactive character of informal channels provides for many of those vital aspects of scientific communication which many scientists currently feel are slipping from their grasp. For example, the *relevance* of information is much more easily established through informal than through formal media. Because of differences in terminology and because there are different fields of endeavour within a science, formal communication is often an inefficient means of providing information necessary for determining the relevance of another's work to one's own. On the other hand, through informal communication a scientist will quickly discover whether he and his colleagues are speaking of the same problems, the same variables, the same concepts, and so on and will guide the exchange to topics of mutual concern and interest. Finally, informal channels enable a scientist to obtain reinforcement and critical feedback which he may wish to receive rather quickly in order to satisfy his uncertainty about some aspect of his scientific behaviour or work. The combination of the requirement that work be well advanced before being reported through formal channels and the long delay typically associated with formal publication tends to render feedback ineffectual when mediated through these channels. In addition, the audience and monitors of the formal channel frequently do not really understand the scientist's objectives because these may not be clearly stated in the formal report.

There is clearly a need for better coordination and more inspired leadership of the communication channels for this important type of exchange of information. Professional societies, sensing this need for better support of informal communication, have experimented with innovations in this

*Note the special problems introduced into the literature by the Proceedings of these Symposia. See, for example, the FID (1962) sponsored study of the "Content, Influence and Value of Scientific Conferences, Papers and Proceedings".

†In view of the importance of informal discussions among peers, etc., many have pointed up the self-defeating frustrations of massive scientific congresses which stifle such contacts by their large size and formality. Thus Moravcsik (1968) has suggested the status-discussion meeting as an antidote to such "super-conferences".

direction. Griffiths and Garvey (1965) have summarized their pioneering activities in connection with the American Psychological Association and more recently in a broader overview of activities including those in optics and physics (1967). Appropos of the APA studies on the significance of informal information exchange regarding "science in the making",* etc., a number of innovations in communication were instituted by various professional societies to make these informal communications more effective. Compare, for example, the Optical Society of America, who in the interest of expediting the transmission of scientific information has experimented with so-called "Long Abstract Papers", beginning with their spring 1967 meeting. Authors who desire to present papers in this manner submit, in addition to the usual 200-word abstract, another longer abstract of no more than four letter-size pages. The larger abstract contains the material normally allotted for contributed papers, enabling the presentation time for these abstracts to be used for a general discussion of the results of the research. The author may use a few minutes at the beginning of the paper for a review of the material and some current results, if appropriate. The total time allotted to the discussion of each Long Abstract Paper is the same as that allotted to each paper in the other contributed sessions. Attendees are expected to have read these abstracts before coming to the meeting. It is hoped that, with proper grouping of such papers relating to similar topics, a strong presider at each of these sessions, and active participation of audience and author, this method of presentation will result in a more direct, active and useful transfer of scientific information than the usual method used currently at scientific meetings.

It is important to note the dynamics in the overall communication process and the ways in which trade-offs and coupling can be made in the various elements of the system. In considering this "dynamics" of the information exchange process, while attempts are being made to expedite the speed of interchange, etc., one must also be alert to the scientific dangers inherent in the speed–accuracy analog of the physical "uncertainty principle" described above. Thus the review and modification process, which is basic to the self-correcting evolutionary refinement of

*This term was first introduced by L. M. McKenzie, in Weil (1954). It refers to nascent science during what we have called the "Eureka" stage of the knowledge cycle (see p. 16).

scientific knowledge, must be accomplished through the expenditure of time; any shortcuts which eliminate the review process incur the risks of degraded quality.

Recent attempts to establish a more systematic organization of these informal channels have come into conflict with the formal publication channels. Something approaching a controversy has occurred in connection with the so-called *Information Exchange Groups* (IEGs). Historically, scientists have always exchanged correspondence and commentaries with their colleagues. It was pointed out earlier that the scientific journal essentially grew out of this practice. However, the IEGs are attempts at formalizing this correspondence for wider groups of workers in similar areas, by means of a central clearinghouse. Beginning very inauspiciously, in February 1961, the National Institutes of Health undertook this clearinghouse role for a number of biological communities. Apparently it became fashionable to be a member of this visible "invisible college", for in 1966 over 1.5 million copies of preprints were sent out! Other fast-moving disciplines desired to emulate this practice (cf., for example, the efforts in high energy physics described by Moravcsik, 1966). One could easily see the process pyramiding into an international explosion. At this point, journal editors became quite rightly concerned with the flooding of this unreviewed, disorderly, uncontrolled material. They took steps to keep this material, once circulated, from entering their journal systems and thereby provoked countercharges from the IEG enthusiasts (cf. Thorpe, 1967 and Green, 1967, for this debate in the field of biology. The latter article contains a useful bibliography of the engendered world debate). Eventually government resources were withdrawn from the central clearinghouse function, thus eliminating these groups as formal entities.

The journal editors have defended their suppressive-actions by pointing out that journals, and their more rapid "letters" counterparts, offer a high-quality and superior channel for the exchange of this type of information;* they are generally critical of the unreviewed nature of preprint material

*The "letters journals", described earlier in this chapter, vary in the amount of review that they receive. In some cases, in order to truly expedite their publication, editors simply exercise an acceptance or rejection review. In other cases, however, the letters are reviewed by editor-selected reviewers in the same manner as journal "papers". In any case, when appearing in journals they receive all the amenities of secondary processing, and are widely available.

and lump the technical report literature together with the IEG correspondence and preliminary writings in their criticism.

While we believe that open expression of scientists among their colleagues is essential, and that freedom of expression and the opportunity for innovation must not be denied to scientists, since it is inherent in the scientific method, we can sympathize with the point of view that wishes to keep down the proliferation of *preliminary* papers and thus to establish enlightened control over the intellectual effluent of scientific production so as to enhance the "Quality of the Technical Environment".

As Dr. Abelson, editor of *Science*, has written (1966): "The explosive growth of the IEGs is in part a mass protest against the inefficiency of many publications. The growth also reflects a desire on the part of some scientists to avoid a discipline essential to the integrity of science."

CHAPTER 6

The Secondary Literature

A. ABSTRACTING AND INDEXING OF TECHNICAL MATERIAL

BECAUSE of the above-mentioned wide scattering of information in the literature, and even more because of the shifting and varied subject requirements of individuals for retrieval, the technical community has become greatly dependent upon the so-called "secondary literature" consisting of abstracts and indexes, as well as upon reviews of literature, for ordered and logical access to the "collective wisdom" in the growing written archive. This dependence has reached the point where "the probability of a paper's being read depends in 80% of cases on the prompt printing of a reliable and accurate abstract of it in a widely circulated abstracting journal" (ICSU: Boutry, 1959). Historically, these secondary media have not always been treated with the support and technical assistance and approbation that their significant role deserves. Authors have often considered abstracting and indexing to be beneath their dignity.* In any case, there has been considerable question as to whether their closeness to the work offers the proper perspective for this undertaking.

However, there are many forces at work in this area which give encouragement that the situation is fast improving. Professional societies are now concentrating their attention and resources on this problem. Journal editors have levied significant title and indexing requirements on authors and have asked reviewers to be attentive to these aspects of the literature.

*It is interesting to note how even book indexing is often relegated to "job-shop subcontractors" whose familiarity with the subject matter is not necessarily thought of as an important prerequisite!

And, on the international scene, the ICSU Abstracting Board has played a useful educational and disciplinary role in establishing standards for editors and authors. In addition, modern technology and the commercial information-handling community have concentrated on this area as an important one and have developed permuted title indexes, mechanized coordinate indexing employing "descriptors" of all descriptions, "citation indexing", which uniquely bypasses any interpretative indexing and directly exploits the author's bibliographic coupling to related works, etc. Also, the Government has provided considerable support for improved information-handling resources within the abstracting and indexing community.

Nevertheless, there are still matters of inertia, as well as some difficult questions and problem areas remaining in this field. Most significant is the intellectual-defying nature of science and technology to be "topic-tagged" (the modern equivalent of "pigeonholing") for any appreciable time. Common-descriptor "thesauri" are usually obsolete when published and the moving front of science and its new ways of looking at things demands a completely up-to-date search strategy from those who would couple to the literature.

Dr. Herschman, in his recent AIP symposium paper on "Indexing of Physics Journals", states the "operational" problem well:

> The problem of indexing physics journals reduces to that of finding reasonable professional criteria for dividing up what physicists do into units of manageable size. The traditional and widely practiced way of dividing up physics was essentially a phenomenologically oriented division, reflecting the belief that the organizing principle for such a division is the theoretical technique used for explaining the phenomena involved. The trouble with this sort of division is that it only superficially and tangentially reflects what physicists do. A much more professional breakdown of physics is 'object' oriented and reflects the belief that physicists consider themselves to be specialists in some type of material or system. The key to the success of such a technique is the determination of the professional divisions and subdivisions of physics. And the key to this determination is the *intimate involvement of professional physicists in all stages of the operation*.

Similar remarks could be made for other disciplines.

The Government has expended considerable resources in making provision for the announcing, indexing, abstracting and secondary distribution of the technical report literature. For the most part, however, such efforts tend to be primarily clearinghouse or what Weinberg has referred to as literature "wholesaling" operations, often without much critical function. It is the concern for quantity and completeness of coverage* which these services often stress, as opposed to filtering and critical review which others point up as the essential ingredients in coping with the information flood, that has led to criticism of the Government's program as being well-intentioned but with the wrong emphasis (Pasternak, 1966).

The professional society publishers of abstracting and indexing systems have been cautious about accepting the technical report literature under their purview. Before they do, they would like to see some "good housekeeping seal of approval" which would certify the reviewed nature of the material, which, of course, is not available.†

Technical reports, once disseminated to clearinghouses, etc., generally receive considerable attention from the standpoint of bibliographic control. Widespread and somewhat redundant announcement, indexing, and abstracting occur. Secondary distribution, either free or at nominal charges, is made to audiences either on the basis of need-to-know or freely to all requestors where possible.‡ Referral services, current effort exchanges, information and analysis services, etc., are all widely supported and are prospering.

As the Government attempts to expedite the provision for secondary services and review products, it comes further into possible competition with the traditional services of professional societies (abstract and indexing services) and commercial publishers (books and monographs). Furthermore, new and promising, primarily government-supported activities, such as specialized information and analysis centers and data

*See, for example, the COSATI (1965) and the antecedent System Development Corporation study on document handling systems (republished as SDC 1967) which recommended that the Government have a policy to collect, *process* and make available at least one copy of all important scientific and technical elements of the worldwide literature.

†A related issue has received a good deal of attention in connection with the establishment of the National Standard Reference Data System at the National Bureau of Standards, involving the problem of arriving at "critical data" certified by a Government "imprimatur". (See, for example, the House Science Committee hearings on the NSRDS, 1965.)

‡Nevertheless, this secondary distribution is not always substantial. Thus, Thompson (1962) notes that "DDC's experience is as follows: There are more than 10 requests for only 10% of the reports deposited with DDC; from one to 10 requests for 80% and no requests for 10%."

reference services, together with new techniques in information reproduction and dissemination, all interact and compete with the traditional society and free enterprise private sector undertakings.

A number of classification techniques are available for processing document information for secondary publications. The fastest and most easily available categorization of a scientific work is through its title. Its significance for classification and filing of the work for subsequent retrieval has been seized upon by even the most primitive information handling systems, to wit, library catalogs. However, with the rising significance of journals over books for the conveyance of state-of-the-art scientific knowledge, these catalogs have become of very limited utility. New, frequently updated "catalogs" of journal article titles have arisen to supply a current awareness and retrieval function. Thus, institutional libraries have gotten into the habit of reproducing the Table of Contents of the journals they receive to widen and make more rapid the awareness of these contents to their constituents. This could greatly expedite the process of familiarization with the literature as compared to the serial passing along of journal issues to the staff, a process that could take years in some organizations and was always subject to infinite impedances in the flow process as interested parties put the journal issue aside for subsequent reading. The alternative system of depending upon readers to scan the journals in the library was also quite unsatisfactory due to the reluctance of most workers to take time away from their "local work environment" for literature perusal.

It was Dr. Eugene Garfield, and his Institute for Scientific Information, who deserves credit, for institutionalizing the reprinting of journal titles in his highly useful series: *Current Contents*. The cooperation of journal publishers to provide advance copies of their tables of contents to ISI allows a very timely listing in these contents journals which are now separately available in Physical, Chemical, and Biological series,* encompassing over a thousand key journals.

However, the great proliferation of journals which has been described above makes even such title scannings rather time-consuming and the specialist-oriented technical worker seeks a more hand-tailored device which cuts across journals and lists works by subject category. This has given rise to machine-produced permuted title indexes, which were

*Available from ISI on a weekly basis.

pioneered by P. Luhn at IBM.* In this operation, all titles are completely entered into the computer, which is then instructed to order and re-order all the significant words in the titles, alphabetically. There are several versions, some of which organize the surrounding words in the title around the indexed word, and are then referred to as KWIC (*key word in context*). Author indexes are also supplied in these same operations. A large number of such services are now available.†

Several other services organize and list the titles in subject categories that conform to a thesaurus of acceptable terms for a particular discipline, or other generally accepted subject categories. See, for example, the British publication of the Institute of Electrical Engineers, which co-operates with the U.S. American Institute of Physics in producing "Current Titles in Physics".

A novel service which includes the concept just described above but which is made available on a quarterly basis on microfiche is called Pandex.‡

As described earlier in the section dealing with journals and their economic plight, the use of these rapid current awareness systems invites selective retrieval based on *local reproduction* of excerpts taken from the accessible holdings of a local or community library. Some equitable solution must be found to properly reimburse or otherwise finance the primary publications which contain this valuable reference material or the system will, as some publishers have warned, "kill the goose that lays the golden egg".

Now, in view of the fact that so much utilization of article titles can easily, and is being, made the question arises as to the intellectual significance of titles for the purpose of conveying information as to article contents. Dr. Weinberg (1963) admonished the technical community to use "meaty" titles. However, the individual author has not customarily

*See his pioneering publication: *KWIC Index*, (1954).

†See, for example, Chemical Titles (ISI); BASIC (AIBS); and the Permuterm Index of ISI which is based on the Source File produced in connection with their Science Citation Index, which is described below. A comparable service for Government Technical Reports, Keyword Index, by OTS, was discontinued in 1963 due to lack of interest.

‡This index is issued quarterly on standard 4 × 6 in. microfiche by Pandex Inc. of N.Y.C. Cf. a review of this service in *Special Libraries* **58,** Dec. 1967.

been charged with technical responsibility for title significance in the technical review of his paper. It is fair to say that journal editors are now beginning to become increasingingly aware of the importance of good titles and reviewers are being asked to critique this part of a work as well as the abstract and full work. Oft-times, other criteria than value to information transfer lay behind a title's choice, and it must also be recognized that the specialist author is not often aware of the extended significance of his work to other areas and is likely to employ very specialized jargon to categorize his work.

We are therefore brought to the recognition that review of a work by secondary processing activities is essential for a deeper analysis of content, and for the wider awareness of the relevance of a scientific work. The *abstracting services* have traditionally borne this important but not highly approbated responsibility. Recently, the specialized information and analysis centers have brought a new and highly significant contribution to this undertaking.

The word abstract is derived from the Latin *abstractus* meaning to draw from or separate out of—. *Webster* defines abstract as "that which comprises or concentrates in itself the essential qualities of a larger thing or of several things". Essentially, the purpose of an abstract is to act as a "surrogate" or replacement for a document wherever possible. It is desired that the abstract could, in some way, convey the basic scope of information contained in the full document, so that others could determine whether a deeper penetration into the document might be germane and useful. There are three basic types of abstracts; descriptive, informative and critical. The descriptive type is usually a statement of the general nature and scope of the document. More valuable is the informative abstract which succinctly informs the reader of the salient features of the work. But the critical abstract is more worthy still in that it further supplies the context of a work and evaluates the contributions that it makes to the advance of knowledge. Obviously this latter effort requires objective skills available only to an expert in the field, who can appreciate the work in perspective.*

There are two other related formats which are highly significant in

*For a good description of the significance of critical abstracts, cf. Juhasz (1965). Dr. Juhasz has pioneered the use of critical abstracts in connection with *Applied Mechanics Reviews*, which he edits.

modern information systems. These are the *extract* and the *digest*. In both cases the abstract length (approximately 200–250 words) is usually exceeded. The first is a shorthand way of enlarging on the descriptive abstract by including material from the report considered especially valuable (notation of contents, key data and figures, etc.) while the second form is essentially meant to be read in lieu of the report.

Who should prepare an abstract? Purportedly, the author knows more about the work he is reporting upon than anyone else (or at least he should)—but can he succinctly describe it? Perhaps not as well as another specialist who will take the time to review the work; but this does take time and effort. It is generally conceded, therefore, that the abstract is the author's responsibility in the first instance, and most journals require that the abstract be submitted with the article for review by the editor and his reviewers. They also supply tutorial material for the author's use to help him in preparing abstracts; material of this type has also been available from national and international professional societies and also from clearinghouses of technical reports.*

The present trend toward further intensive use of machine methods in information handling has placed added significance on the abstract, just as we saw the earlier stress that was brought to bear on the significance of titles. With the increased memory capacity of computers, it is now possible to store and process the abstract in computer search operations, so that the terminology of the abstract becomes of great significance in determining whether the material is retrieved as relevant or not. If the abstract expresses the content of the document badly or inadequately the whole document and its potentially significant contents may be lost from sight by those using the information service, for as we have seen earlier it is by reference to abstract tools that most literature is retrieved.

Another technique for describing the content of a work is that of key terms, words, or other descriptions of content. This technique for ordered entry into the contents of a work is called indexing and is of course already widely familiar to all those who have perused books. The need for commonality in language for applying "descriptors" has led to the elaborate organization of "thesauri" or dictionaries by fields and disciplines. While this work has customarily been handled by lexicographers who started

*Cf., for example, the AIP's guide, the ISCU abstracting board's guide, the DDC guide (AD 667 000, March 1968), etc.

their efforts in connection with hierarchical organizations of knowledge so as to arrange books in some serial order on linear library shelves,* it has now become recognized as a scientifically demanding effort by professional societies. There have therefore been a rash of efforts to systematize the "descriptor" language in the Engineering, Physical, Chemical and Report literature, areas (Project LEX, the largest of these efforts, has recently been completed by a DoD-wide group (1968)). However, a growing and dynamic science will require frequent updating and rearrangements as new insight into the science itself alters the intellectual organization of the technical subject.

The application of descriptors based upon a universally accepted standard terminology will certainly assist in the organization of technical material for retrieval and such an approach is currently being pressed on authors and their societies. The DoD now requires attachment of a so-called form 1473 to technical reports, giving an abstract and assigning descriptors in addition to other vital information concerning a work. This can then be uniformly processed by various readers and secondary groups into information handling systems. COSATI has recently recommended (1967) a guideline for a government-wide equivalent of this form, to be utilized as a title-page for the report itself, thus giving further consistency to report format.

The availability of descriptor terms for a report has led to the development of various techniques for using mechanical systems and search strategies to identify works for retrieval. These come under the subject heading of "coordinate indexing", pioneered by the late Mortimer Taube (1953). Essentially, these techniques allow the retrieval of reports that include the intersection or commonality of two or more indexing terms. In the case of modern computer systems these additionally permit the use of Boolean logic searches which may add further to the selectivity of pertinent works (e.g. all reports in a system dealing with commonality of descriptors A, B and C and excluding D). A number of computer systems, using the recently developed powerful techniques of time-shared, real-time access to computer information banks, allow the retrieval of pertinent documents or journal articles based on the direct search strategy of the

*The custodians of the Dewey Decimal System and its international extension the UDC (FID) and the Library of Congress system were pioneers in this activity.

user, bypassing a library intermediary or printed index tools and thus eliminating difficulties previously encountered in librarian communication problems, time delays, or out-of-date compendia. With the advent of computer coupled-visual displays (including interactive ones) and rapid printouts, we have achieved a whole new revolution in the retrieval field. These techniques will be described at greater length in Chapter 8.

Mention should be made of a unique retrieval tool which adds a new dimension to the search for relevant references. This is the Science Citation Index, available from the Institute for Scientific Information. Its feasibility is dependent upon two basic phenomena; one is the strong tradition within the scientific literature to explicitly reference the antecedent literature that is significant to its development and on which it is dependent for greater detail of explication.* The other is the observation that references can easily be noted by untrained scanners and processed by computer to "invert the matrix" of coupling to the citing works. The SCI is prepared by performing this inversion and thereby expeditiously allows the retrieval of any and all works (within the corpus of screened journals) which refer to any subject work. This gives the illusion of time reversal since it provides references that are later in time than the work referenced, but of course it is still necessarily backward in time from the publication dates covered!

The great significance of the SCI is that it enables the selective retrieval of literature without any of the customary intrusions of previously described secondary intellectual processing, with their inevitable distortions. There have been a number of literature search comparisons between this tool and the more traditional tools and the conclusions show that each seems to offer distinct advantages of a complementary nature.†

A current awareness service is available from ISI based on common citation coupling (ASCA) which also has analogous advantages for staying

*The use of frequency of citations to a work to somehow achieve a measure of its significance are inevitable but must be handled with care if they are not to lead to distortion.

†In addition to the use of this tool for the *a priori* measure of a work's significance through the amount of citation in peer works, a number of practitioners of the budding "science of science" have utilized SCI as a technique for identifying "paper networks" so as to trace the evolution of a theory or experimental technique and its diffusion into the collective intelligence. Cf., for example, Price (1967).

alert to relevant works which may fall outside of one's immediate specialty or literature scanning.*

Re-emphasis in the U.S. on the use of abstracts and indexes, and of their support, coordination and mechanization, occurred soon after the phenomenon of Sputnik which forced the U.S. scientific community to search itself for adequacy in meeting the Soviet challenge. Publicity concerning a centralized S.U. approach to information handling in their All Union Institute in Moscow, known as VINITI, which publishes a mammoth multivolume abstract journal (*Referativnyi Zhurnal*), caused the U.S. to question its traditional plural-society arrangements. Fortunately, cool heads prevailed and no "mad momentum" to emulate the Soviet system (which has since been characterized by a Russian scientist as all "input and no output"!) took place. However, a greater recognition of the importance of A and I activities in the U.S. community led to a coordinative assembly, the National Federation of Abstracting and Indexing Services (NFSAIS), and perhaps more significantly, a greater investment of funds and dedication of professional societies to the improvement of the quality and timeliness of these services. A number of guides to the world's services in these fields have been compiled† and the organizations themselves have been re-dedicated to useful service.

Nevertheless, the amount of duplication, the need for coordination, and the issues surrounding the techniques for support of these services have continued to pose problems and to defy solution. COSATI contracted with the System Development Corporation for a systems analysis of this problem and the latter's findings, reported in *A System Study of Abstracting and Indexing in the U.S.* (1966), are significant:

> The present ad hoc "system" for abstracting and indexing in science and technology is beset by a number of problems shared by three major parties: the

---

*This significance of interdisciplinary communication deserves greater attention in its own right. Redundant publications in different discipline channels and repackaging as in technological utilization efforts are important in this regard. Popular interdisciplinary journals which present state-of-the-art reviews and a greater concern for heuristic publications emphasizing questions requiring further investigation would be helpful.

†See, for example, the NFSAIS sponsored Library of Congress (Science and Technology Division) publication: *A Guide to the World's Abstracting and Indexing Services in Science and Technology* (1963) and a companion publication by the FID (1964).

> users, the A & I services, and the Government. The system also has many limitations, both with respect to current needs and to future requirements. The problems of the various system elements interact with each other in sufficiently complex ways to preclude easy or immediate solutions.

A number of specific recommendations to COSATI to attempt to rectify this situation are under review.

The difficulty which the plural U.S. system faces by virtue of the multiplicity of interests was apparent in a conference that SATCOM, the National Academy of Sciences' scientific information advisory committee to COSATI, recently called, which was attended by representatives of the various elements of the "private sector". It would appear that the marketplace will have to be depended upon more for an expression of the satisfaction of user needs, and the professional societies, commercial interests and specialized information services will have to compete for this satisfaction of the "customer". However, it appears that the user's purchasing power for these needs will still have to be provided by the ultimate user, namely the public through its governmental agencies who are pressing for the advances from science which are served by effective scientific communication techniques. This technique is preferable to the direct subsidy of secondary services by Government which is open to the danger that such services will develop independent of the user's real needs.

One important aspect of this economic issue has been noted by Loosjes (1967), who points out that "abstracts journals mostly end up in non-commercial channels and have to fight against financial troubles, whereas *surveys* of the literature (as described in our next chapter) tend to be self supporting". He therefore advises learned societies to adopt the latter form of publication! As will be seen later the same situation arises in connection with the services of specialized information–analysis centers, who have also begun to avail themselves of the commercial publication route for some of their secondary products and reviews.

Professional societies have attempted to pass some of their extensive information handling costs along to the *sponsor* of the research reported upon (on the basis that a research effort is not completed until it is both published and disseminated and thereby recognized and commented upon), via the author. The page charge for publication is the prime example of this. Now, it is proposed that this principle be extended to the secondary dissemination of information about a work (by virtue of its significance in bringing the work before the community, as mentioned earlier in discussing the significance of abstracts and indexes to the coupling between authors and users), with an additional abstracting and indexing charge to the author. Carried to its extreme this line of argument would make the cost of libraries, information centers, and even the provision of time for the reading of materials a responsibility of the author who has the effrontery to publish! Clearly this *reductio ad absurdum* makes it apparent that another approach must be used which puts more of the burden of costs on the *user*. However, it must be admitted that this still places the public, and its agent the Government, in the position of sponsor since the user of information is just another permutation of the author universe. Essentially the whole enterprise of research has come to be considered a worthy grantee of public funds. As long as this is the case, it is worth considering how the overall information communication process can be improved in the most cost-effective manner. This should take advantage of the skills, capabilities and resources of all phases of the Government in-house, professional society and commercial sectors. However, it is recognized that to raise these issues is not to resolve them; much thought and coordination will have to be given to the proper combination of these efforts on a national and international basis for an efficient and viable information system.

B. REVIEWS, MONOGRAPHS AND STATE-OF-THE-ART REPORTS

It is generally recognized that the critical state-of-the-art review potentially provides one of the most significant contributions to the advancement and diffusion of knowledge. It is this technique, which we characterize as "scholarship", which permits the selection and compaction of knowledge, as well as its restructuring, and thereby makes it feasible for new scientific workers to assimilate previous efforts and build upon the shoulders of their predecessors.*

*A number of authors have drawn upon a botanical analogue in viewing the "hit or miss" dissemination of information through journals as being on a par with the primitive random process of wind-pollination in plants, whereas state-of-the-art reports correspond to the technique of insect-pollination where, with less pollen, more get to the right flowers (after Scott, 1966).

The Weinberg Report (1963) pointed out that scholarly and critical reviews and similar publications can play a major role in easing the information explosion by condensing and summarizing information from many predecessor documents. Price (1964) notes that this effort of scholarship is the technique by which the educational system copes with the dichotomy between the mounting archive of knowledge and the fixed period for education and assimilation of this knowledge prologue.

> Thus, while the number of papers produced per scientist is still the same, approximately as it was in the 17th century, and though there are more people involved, by many orders of magnitude, we have divided and ruled by the trick of specialization. Each specialty grows exponentially, so as to double every decade or so, just as it did in the days of Newton and of Franklin. We solve the problem of mounting literature in each specialty by a curious expedient called scholarship. This is the art of packing down the accrued knowledge through more and more economical statement so that eventually it becomes part of the material that can be learned by the student before he arrives at the research front.*

Nevertheless, in spite of the general approbation of the scientific community for this noble undertaking, there is a feeling that reviews are not being written in the numbers and quality required, and there is difficulty attracting qualified individuals for these formidable tasks, as well as in wisely spending monies even when earmarked for this purpose by government and professional societies.†

*It is interesting to note that should this function get too far behind, the information flood would recede on its own since the new generation would never get to the research front to make its contribution! In this sense there are some "stabilizing" elements which prevent a true "explosion" in the information growth process. On the other side of this argument, one should note that "positive feedback" as introduced by better information practices can speed up the information growth process in the course of making greater progress. This dynamic of the information chain reaction process is not often raised in the discussions of improved information handling through mechanization, etc., and we are afraid that the scientific community has not yet realized just what is truly in store for it in the future.

†See, for example, the Weinberg PSAC Report (1963) and the recent *Physics Today* article by Goudsmit on this subject (1966).

There is even some question as to whether the literature is worth retrieving (Goudsmit, 1966) or reviewing (Branscomb, 1968): it is therefore the first responsibility of the reviewer to exercise great selectivity.

There are a large variety of *review* media which help the technical individual to cope with the information flood. In the order of increasing complexity and preparation effort, review materials include:

1. *The announcement bulletin.* Here the reviewing group has reduced the "entropy" or disorder of the total literature by sorting out material conforming to a certain category of selection. Specialist journals, clearinghouse bulletins, publishers' lists, specialized information center accession lists, etc., provide such material. This category can appear on a periodic basis or be prepared *ad hoc* to satisfy a search request, in which case it is called a demand bibliography.

2. *Annotated lists.* This format adds some information to that conveyed by the mere announcement, in that it usually gives some added notation of the content of materials beyond author, title and subject. If the annotation is done by an expert in the field, such as in the preparation of critical abstracts or reviews, then the material takes on an especially valuable guideline function. Primary journals and abstract journals often include material of this type in a special section as a distinct service to their subscribers.

3. *Review journals.* These are published in various fields and represent state-of-the-art compendia for the journal literature, analogous to that described above for the report literature. Mention has been made of the approbation which such works deserve, however, the demands on authors in such preparation is exceedingly great. However, as is pointed out by Goudsmit (1966), the rewards are also great in that the definitive review of a subject is likely to be remembered and referenced by subsequent authors preferentially over the source works themselves.

4. *Encyclopedias*, separately available for individual technical disciplines, are becoming increasingly popular as a kind of shorthand guide to the subject matter in a field.* The various entries in such tomes are usually prepared by separate authors who in essence have written concise "reviews" of their subject, often with liberal referrals to the primary source

*See, for example, the Reinhold family of one-volume encyclopedias in the Physical, Chemical and Biological fields and the Larousse series on a more popular level.

literature. Composite science encyclopedias are also more widely available at different levels of sophistication.*

5. *State-of-the-art reports*, as issued by individuals or information centers (as described below), are now recognized as a significant contribution to research in their own right. Often, data compilations, as they may be prepared by data centers (such as those under the general cognizance of the National Standard Reference Data System (q.v.)), offer a new perspective on physical phenomena which reorients the scientific perspective in a particular area. Weinberg (1963) noted the significance of this type of effort in his description of the work leading to the discovery of the "shell model of nuclear structure" which arose from a data review of nuclear properties by the Nuclear Data Center.

6. *Annual Reviews*. This format is becoming exceedingly popular in many technical areas as a technique for periodic preparation of state-of-the-art reviews of important dynamic research fronts. These are usually published in a series of subject reviews bound together in monograph form. This format seems to command greater approbation as well as effort on the part of the writers and has further advantages of conspicuity to the researcher. Annual Reviews, Inc. (Palo Alto, California) has pioneered in this area. A large number of publishers are now active in producing such reviews in different fields.

7. The *full monograph* or book may be devoted to a topical review of a particular area. Historically this has been the primary review medium and will be further discussed below; however, it should be noted that because of its great comprehensiveness it usually is delayed in time of availability to the point where it is of less significance to the growing edge of research than it is to tutorial applications involving wider diffusion of knowledge.

8. Lastly, there are a number of what might be referred to as *tertiary materials*—such as *review* of *reviews* or *bibliography* of *bibliographies* which facilitate a wide screening of review materials, so that subject coverage can be identified.

*See, for example, the *Handbuch der Physik*, *The Pergamon Encyclopedic Dictionary of Science and Technology*, *The McGraw-Hill Encyclopedia of Science and Technology* and other more exoteric works such as the World Book Science Series, *Harper's Science Encyclopedia*, etc.

As indicated in the earlier discussion of the widespread generation and use of *technical reports*, there is an especially critical need for the process of filtering and review of this generally poorly refereed product. One technique, suggested by Dr. Weinberg (1963), to establish a government "screening agent" on the contractor's premises for this review, has not found an enthusiastic reception. Of the various possibilities, it would appear that the specialized information and analysis centers, with their qualified staffs and continued attention to the subject matter, offer the greatest reviewing potential. Their state-of-the-art reports and annotated bibliographies offer a useful service for calling worthwhile developments to the attention of those seeking reliable information in their fields of specialization.

Fortunately, this significance of specialized information and analysis centers* for digesting, refining, and "compressing" the report literature through "state-of-the-art" survey reports, is receiving widespread notice and approbation. Battelle Memorial Institute, a pioneer operator of such centers, has recently made a survey of policies and guidelines followed by these centers in the preparation of state-of-the-art reports (Darby and Veazie, 1968). They use as a working definition: "A state-of-the-art report is a comprehensive analysis of available knowledge (published and unpublished) on the status of a particular subject area or mission, frequently written for the use of a specific reader audience." In this process of identifying the technical audience and their needs, as well as in the critical functions of selecting a subject that is ripe for a review, the selection of a team of authors, and in the provision for an information base of relevant literature, the information and analysis center is ideally and strategically located.

The process of extraction of the valuable contributions of the primary literature is of course exacerbated by the verboseness and imprecision of so much of this literature. This has been criticized by PSAC member Dr. Branscomb (1968) who points out that it may be best for the reviewer to approach the problem *de novo* and do the critical research himself! Perhaps this is too much to ask for and we would hope that information can be so organized as to make life somewhat less exasperating for reviewers. Dr. Kay Way, in her charming paper, "Waiting for Mr. Know-It-All" (1962), points out that good author practices of title choice, abstract

*For a general discussion of the rising development of the concept of information and analysis centers in identifiable mission and discipline oriented areas, cf. Chapter 7.

writing, assignment of descriptors and, of course, improved writing style, could do much to enable data centers to do their job more efficiently.

Neal (1967) has a very interesting innovative suggestion:

> I propose that the author of each new scientific and technical article or other publication append to it a separate statement of the "Essence" of each new or improved "item" of knowledge that the publication contributes to the social body of scientific and technical knowledge. . . . These essences would be co-ordinated by a group (through professional societies, etc.) and integrated into a "viable set of Scientific and Technical Organized Reference Essences" (STORE)!

Oh, that this were feasible; reviews would be self-generating!

Particular mention has been made of the *compaction* function of review materials over the more extended and diverse primary literature. This aspect of the review material has recently been emphasized by Dr. C. Sherwin in his prospectus on a global network of mechanized information systems (1968). Because of the capacity limitation of even the largest computer memory systems, it is necessary to get approximately a 100:1 compression ratio from the original periodical and report literature volume, if automated information storage banks are to be practical. Dr. Sherwin estimates that this is precisely the factor offered by critical state-of-the-art works over their reviewed, and presumably then retired predecessors. Unfortunately, I believe that state-of-the-art reviews, as valuable as they are, do not really obsolete their precedent primary works, but rather *select* the significant elements of this literature which are perforce referenced and will have to be referred to for perusal by the specialist who seeks to benefit from their more detailed discussion and data. The truly definitive review which contains all the essential aspects of its predecessor works and makes their referral unnecessary is a "rara avis" indeed.

Numerous recommendations have been made for expanding the number and quality of reviews (e.g. Forney, 1963; Licklider *et al.*, 1965; Goudsmit, 1966). The opportunity that professional societies and government support could offer in this regard, has been suggested but not yet fully exploited. It is true that the community support of information-analysis centers has greatly assisted this undertaking, and it is hoped that this area can grow in the future. Unfortunately there is some confusion of the role of the Government in this regard. Dr. Adkinson (head of NSF's OSIS) has recently stated (1967) that "this (area of critical reviews) is only symtomatic of a basic lack of agreement on the division of responsibility for scientific information between private and government sectors of the economy . . . I might point out that this problem is an order of magnitude more vexing when considered from an international point of view."

Here, again, the dangers of possible control of the objectivity of this important scholarship function can be minimized by channeling the support through specialist centers which can better judge the experts in the field and can insulate them from the establishment's technical "party line". However, there are cases, such as recent review texts sponsored by AEC and NASA, where direct support has also worked admirably to provide the community with important survey resources.

Of course, the question of quality can be raised with regard to the review function just as to the primary materials themselves. The significance and responsibility of a reviewer make this a task for a "master". This is why selection by a peer group, such as a professional society or specialist center, is so important. Naturally, those whose works are adversely criticized or who are given short shrift in the review are likely to be offended. This issue has been raised in connection with the pending legislation attempting to formalize the National Standard Reference Data System (NSRDS) set up by FCST direction at the National Bureau of Standards, where it is proposed that critical standard data be given an "imprimatur". As was pointed out by Dr. Seitz,*

> the use of a special symbol could create two problems. First, the Government, by adopting a mark of approval, may expose itself to embarrassment when erroneous or inaccurate data find their way into the data compilations. Second, scientific data are subject to constant improvement; debate about the correctness of data is a continuing process. Using a mark or seal of approval on the data may tend to freeze or impede the improvement process.

In the final analysis the credibility of any critical review will be dependent on the credentials and expertise of the reviewers.

With regard to the most formal review materials, namely books, we have it on good authority that "of the making of many books there is no end". The worldwide commercial publishers currently issue some 20,000 new technical titles a year; 6000 in the United States, alone. These have become profitable ventures for the publishers, whose numbers also have

*In testimony before the House Subcommittee on Science, Research and Development concerning H.R. 15638 on NSRDS legislation.

increased greatly, especially because of the ready market from libraries and the knowledge-hungry technical community. Nevertheless, there is a general feeling that technical books are not "better than ever". We believe that this is often due to the difficulty of attracting outstandingly competent individuals to these time-consuming and personally demanding tasks. Since the review or book becomes a community resource when completed, we believe that it should be treated as such in its preparation, deserving of the same community support which other less neglected elements of the communication system are receiving.

There is, however, an asymmetry in the process. Whereas the completed book becomes a technical resource, to be utilized and exploited as an official part of the responsibilities of technical staff members and purchased and retained within the official information center facilities, all with perfectly reasonable justification since it is a most significant element of the information system, nevertheless, the investment of the time, effort and resources into the preparation of the book is normally considered to be worthy only of extracurricular activity. Thus books are not even given the institutional support available to the very minimal technical report! They are produced in the "spare time" of the author usually at the sacrifice of his evenings and weekends. Of necessity, the spare-time approach to such writing must delay the culmination of the task to the point where books are notoriously slow in appearing and are therefore not nearly as useful for the conveyance of up-to-date information as they should or could be.

As far as the economics of book preparation are concerned, the work is economically advantageous only to the publisher; the author rarely gets a minimum wage for his time and effort. One might say that this is beneficial to the ultimate quality control of the enterprise, since it must perforce attract only dedicated workers rather than profit seekers. (The less money, the more honor as the Quaker proverb goes.) I would argue, however, that the first rank of experts who are really ideally qualified for such rigorous undertakings are not necessarily attracted to these tasks by the present system. In fact, since the publishers stand to gain the most in the entire operation, it is to their advantage to promote a greater number of average texts rather than a smaller number of excellent ones. Furthermore, the stress on completeness, rigor of presentation and printing of all needed supporting figures, etc., which is desirable from an information standpoint, runs contrary to the economics of the venture, so that some compromise is insisted upon.

This is an area where government sponsors, professional societies, information and analysis centers, university presses, etc., can all do much more to enhance the quality of the technical environment. For example, there is no reason why a government sponsor, mindful of the great potential significance of a monograph text as a resource in a fermenting area of science or technology, cannot commission a work by a peer—recognized "master"—in the field and provide him with the resources to complete a quality effort in a timely fashion. This work can then be distributed in a number of ways to the satisfaction of the various interests concerned. (The Atomic Energy Commission has pioneered in this approach.)

Clearly the specialized information-analysis centers have much to offer here, in the choice of authors, selection of technical subjects ripe for such treatment and in the information support necessary to make the work converge in a timely fashion. Furthermore, the channeling of support through such a center can de-couple the effort from the perhaps overly subjective constraints that might obtain by administering the effort directly from a government sponsoring agency.

Having supported the preparation of the work, the sponsor is then able to have the work published for a very low price since he seeks to make the work widely available so that other R and D efforts might have the benefit of this up-to-date, coherently packaged information. This enables the work to be sold at a reasonable cost, so that an ample number of copies may be purchased and placed close to the work environment of practicing workers, which must be compared to the rather inefficient limitations posed by very expensive commercial products which seem to price themselves out of the individual market and thereby be available only through the cumbersome process of inter-library loan.*

Another practice in which timely publication of a quality product is supported, is through the mechanism of alternate publication in book form of a work submitted in lieu of a final technical report in satisfaction of a contract. This is especially popular in connection with efforts in the

*A number of examples of texts that conform to this suggested practice come to mind, e.g. the ARPA-ONR-IRIA work: *Handbook of Military Infrared Information*, and the NASA *Space Handbook*.

foreign affairs, political science and behavioral sciences where the technical report format has not achieved widespread recognition and is likely to be missed by the community. Sponsored works of the "think tanks" and university scholars are often published in this form, and makes for a very useful product, one which clearly serves the best interests of the sponsor as well.

Let us now look at the issues involving the announcing, cataloging and indexing and abstracting of the book literature.

The publishers have historically dealt with the *librarians* as their point of contact. Their "flyers" and announcements, as well as the secondary publishing trade journal literature, are designed with them in mind for an audience. As we have seen earlier, the traditional library approach to technical literature leaves a great deal to be desired and the bridging of the gap between technical staffs and their libraries seems to be a serious problem. Librarians are really quite at a loss in this effort to select meaningful book purchases in advance of requests from their technical staffs and before technical reviews of the material appear. Nevertheless they are forced to make such selections so as to at least give the appearance of service. One therefore finds that at many important technical libraries the holdings of the library are as much as 80% determined by the technical choices of librarians and other "nontechnical" personnel. Obviously they are swayed by what they consider to be significant titles and subjects as well as the advertising glitter sent them by publishers. Needless to say, this rather routine purchasing of all and sundry technical products of the publishing world feeds back a reduced incentive for quality in the first instance and promotes the quantity of titles which contributes to the information flood.

There is a further aspect of the book literature handling which contributes to the reduced significance of what could be a much more important medium for scientific communication. This is the shortcoming in the indexing and abstracting function which is so significant for further retrieval and effective knowledgeability of the contents.

First of all, the actual indexing of material contained in the book is usually performed by nonspecialist personnel! This is an historical custom which should be rectified in the same way that indexing has been upgraded for journals and technical reports. Utilization of community accepted descriptor-thesauri, etc., should be encouraged, but here again the problem stems from the economics of the process and could be improved in the same manner that has been suggested above for the support of a higher quality product.*

Moving on to the classification process, we note that the assignment of the book to a specific lineal position on the library shelves has been the principal concern of librarians, instead of the efficient coupling to the minds of present and future users of the library! In conformance with this desire, and without the necessary technical staff, the librarians have become dependent upon the Library of Congress for a catalog card service. While this does make for uniformity it leaves a lot to be desired as far as technical information handling is concerned. Thus, as a critical government task force has noted:

> within the broad scope of the Library's responsibility and interest, all fields of knowledge—music, art, history, science, etc.,—are of essentially equivalent importance. If different degrees of importance do exist and affect priorities of processing, these are likely to result from the relative interest and demands of the thousands of libraries which rely on "LC" for catalog cards and other service. To many of these libraries, a book on Howdy Doody is more important than a document concerning astrophysics, because of the demands of their individual users. Processing within LC must naturally be responsive to such differences in demand (Crawford *et al.*, 1962).

Even after the delay in receiving the LC catalog cards, what the library catalog system contains is only the traditional Author and Title Index system with some additional subject entrics that are usually based upon the title terms. What about the amenities of abstracting and descriptor assignments that are given to journal and report documents? Unfortunately these are usually omitted for this significant medium! There are a few specialized services which try to provide additional abstracting for books especially where they contain a number of separate topic chapters such as, for example, in the case of the symposia proceedings and annual reviews, which are of growing popularity in book format. The International Aerospace Abstracts of the AIAA, sponsored by NASA, is particularly good in this regard.

The book literature does have an important quality control element known as a book review which is often included as a supplementary

*There has been a suggestion to use book-indexes as building blocks for a cumulative index (Kochen, 1967) but this must be done with great selectivity and care.

section in journals. These reviews, unfortunately, appear so late after publication of a book that they can hardly serve to assist libraries in their acquisition function. Nevertheless, they do serve to acquaint many people with the availability of a book. Such reviews are rarely critical to the extent that a book is panned. (Perhaps this is a function of the general phenomenon of the scientific fraternity taking in each other's washing.) However, they are useful in delineating the subject matter of a book in much greater depth than can be gleaned from the title or the dust jacket "propaganda". A useful, but not widely known, reference tool is the SLA's *Technical Book Index* which brings together excerpts of such reviews (which appear in the popular or learned journals) in a monthly compilation. There is also the analogous Book Digest, primarily for the nonspecialist book literature. An analogous British publication is the *Technical Book Review* (TBR). This general format is an example of what might be called tertiary literature, i.e. review of reviews.* There are also bibliographies of bibliographies, and general *guides* to the literature (e.g. the John Wiley series which covers the space literature, etc.). Some of these are published by library groups, professional societies, etc., and others are commercially published by literature handling houses such as Bowker (Books in Print), Wilson, etc. Pergamon Press has a series for Organic Chemistry, known as "Index to Reviews and Monographs in Organic Chemistry" about which they say:

> . . . will allow research workers, teachers and students to locate quickly those current reviews which may be pertinent to their work. The great majority of the articles and books cited in this issue are fairly complete, authoritative reviews and monographs which are particularly useful as a first step in assessing the total original literature.

Mention should also be made of the encyclopedia year books which contain good tutorial compilations of reviews in different fields. The McGraw-Hill *Yearbook of Science and Technology* is especially valuable in this regard.

Mention has been made of the growing book literature format of symposium proceedings and annual reviews. In a sense these are "non-books" because they usually lack any kind of homogeneity or coherence,

*Loosjes (1967) refers to these materials as "abstracts journals squared", "that is to say a second collection of the same materials for inquirers who are not accurately enough served in their specialized field by more general abstracts journals".

which has heretofore been the hallmark of the monograph's significance to the analysis and digesting function. Nevertheless, they do have some connectivity, either through the choice of papers for a symposium or by the Editor of the *Annual Review*. One might question whether there is advantage in publishing this variegated material under one cover, rather than in individual documents, recognizing the greater depth of indexing and abstracting that they might receive in the latter category. Nevertheless, the historic approbation, real and imaginary, which the "hard-cover book format" seems to bring to all involved with the effort, and the commercial incentives to publishers to proliferate this format seems to contribute to their increased incidence. However, it is my opinion that these works have a very short half life of significance, and time should be of the essence in their publication. Unfortunately, unless the discipline offered in the process of the organizational effort is used to converge rapidly on a product, these efforts have a way of being delayed by authors and publishers which makes the work obsolete when published. There also tends to be a great deal of redundancy in the various products of such symposia which further proliferates the various archives. Thus we have preprints, documents and papers, journal articles, reprints, special journal symposia proceedings, all appearing before the ultimate book which publishes the symposium proceedings.

As has been pointed out, there are two parallel functions to the role of information, those of the increase and diffusion of knowledge. In serving this latter function the book format is really supreme. It is in this use as a "text" for seminars and technical courses that the tutorial format of the book finds its origin. As was indicated in the earlier description of the "knowledge cycle", the frontier understanding of the phenomena of a field must finally be packed down through the function of scholarship to become part and parcel of the "collective intelligence" of new workers in the community. This part of the communication system is handled by the University System rather than professional societies or other phases of the industrial and governmental community. However, it should be pointed out that the government support of university research has provided the principal, albeit indirect, supporting base for the training and education of graduate students. To this degree the Government can be and has been concerned about the literature available for this tutorial effort. In addition, the Government through other educational auspices under NSF and

HEW has been involved with many other aspects of the educational curriculum. It has taken the initiative to advance the techniques of basic presentation of the sciences and has commissioned new textbooks and course manuals in physics, mathematics and biology. These new approaches have been credited with radically upgrading the quality of instruction in secondary schools and colleges. The manner in which these textual materials should be made available has been a matter of concern to the Government and to commercial publishers and has led to a debate concerning the related copyright practices. (See Marke's *Copyright and Intellectual Property* (1968) for a balanced discussion.)

In summary, it is ironic that the monograph literature, which potentially represents the zenith in terms of the high quality and intellectual resources for scientific information handling, seems to attract the least attention in terms of community concern for its improvement and in the details of the processes of author selection, sponsorship, publication quality and timeliness, current awareness tools, indexing and intellectual processing, distribution, handling and retrieval.

Clearly the entire area of the book literature requires a great deal more attention from the technical community in view of its great significance in the total technical communication system.

CHAPTER 7

Specialized Information and Analysis Centers

ONE technique, by which science has attempted to cope with the rising flood of information, which results from the growth of scientific endeavor and the so-called "second Malthusian dilemma" (the n^2 relationship which results from the far greater number of interactions, i.e. $n \times n$ that result from an increasing number n of scientists, as well as the dilemma of increased complexity)* is that of a *division of labor* between those who create or discover the facts of science, and those who sift, absorb, and correlate the facts. The scientists and their technical information assistants who perform the latter functions have come to be known as constituting "specialized information and analysis centers".

The explicit formulation of such centers may be a relatively recent institution, however, the concept of individual and organizational experts available for consultation is a very old cultural device. Thus, as Dr. E. Bering (of the National Institutes of Health) noted, in remarks at the National Bureau of Standards symposium on such centers (COSATI, 1968):

*Realization of this aspect of big science is credited by Dr. Weinberg (1967) in his charming book, *Reflections on Big Science*, to his mentor Professor Eugene Wigner. In the latter's 1950 essay on "The Limits of Science" (Wigner, 1950) he pointed out how this n^2 singularity would eventually place limits on science which would result in science undergoing a social reorganization consisting of a layering into several hierarchies, varying from bench scientists up the layers of the structure to the philosophers at the top who "know almost nothing about almost everything".

> The specialized information analysis center is almost an inherent universal in the behavior of man. They have existed almost as long as man has. In Greece, it was the oracle at Delphi. In primitive tribes, it was the medicine man. And as history went along, the information centers were the monasteries during the Dark Ages. There were the Universities, then after Columbus, charts were needed and the British Hydrographic Office developed. . . . This is a natural, social development of man.

However, there is no question that the recent growth of science and technology has greatly stimulated the requirement for formal specialization and community cooperation in coping with the information deluge. The concept of such a center in recent times has evolved from the need of individual workers for a staff of technical information assistants, beyond the realm of practicality for individuals, but available collectively to the community, to serve as a middleman or retailer to screen the literature issuing from the multitude of "publishers" and information wholesalers so as to provide a more personally useful product to the technical consumer.*

A recent COSATI Panel on this subject (1968), has adopted the following definition of an Information-Analysis Center:

> A formally structured organizational unit specifically (but not necessarily exclusively) established for the purpose of acquiring, selecting, storing, retrieving, evaluating, analyzing, and synthesizing a body of information and/or data in a clearly defined specialized field or pertaining to a specified mission with the intent of compiling, digesting, repackaging, or otherwise organizing and presenting pertinent information and/or data in a form most authoritative, timely, and useful to a society of peers and management.

In the "bandwagon move" to organize for information handling, a large number of library-like activities have changed their names to information centers, but the distinction which characterizes an information-analysis center must not be overlooked. As stated aptly by Weinberg (1963):

> We believe that the specialized information center should be primarily a technical institute rather than a technical library. It must be led by professional

*The PSAC report, *Science, Government and Information* (1963), emphasized this economic analog to the normal consumer-retailer-wholesaler relationship. While the recent great growth of specialized information-analysis centers dates back to World War II, and the Big Science period, and a useful NSF directory of such centers with descriptions of the concept was published as early as 1958, this "Weinberg report" gave great impetus to the concept because of the official approbation which it represented, and such centers have proliferated and prospered ever since.

> working scientists and engineers who maintain the closest contact with their technical professions and who, by being near the data, can make new syntheses that are denied those who do not have all the data at their fingertips. Information centers ought to be set up where science and technology flourish.

COSATI (1968) has identified at least 113 *Federally supported* information-analysis centers, which serve in mission or discipline oriented areas, from Aeronomy and Space Data Services to X-Ray Attenuation Coefficient Information Center. The 1958 NSF directory identified some 400 private and government centers and the Library of Congress' National Referral Center for Science and Technology (set up by NSF grant to keep the roster current) has descriptive literature on several thousand significant institutional information resources in Science and Technology in the U.S. alone (LC, 1965).

Clearly the information-analysis centers offer a great opportunity, for both advancing the state of knowledge and providing data and information syntheses and reviews as well as in providing services to individual workers in the form of bibliographic assistance and in information retrieval. (Actually most information centers provide detailed information responses together with referrals to the cited literature; the basic source materials referenced must be obtained through publishers or libraries.)

A number of areas where these contributions are significant have already been mentioned, such as in *reviewing* and critiquing significant literature that is issued in technical reports, providing literature guides and annotated bibliographies, as well as in the most significant area of providing state-of-the-art reviews.

For a recent comprehensive review of how centers go about preparing reviews, etc., cf. Darby and Veazie (1968). Their survey finds that the follow-up contacts subsequent to preparation and distribution of such survey reports are one of the most stimulating and vital aspects of state-of-the-art writing:

> The user audience may react by providing additional information or data which may not have been previously available. Further on the basis of their experience or knowledge, they may be able to further clarify or apply the principles discussed in the report. If a revised version of the report should become desirable, as it might in a rapidly developing technology, a broader information and data base may thus be available.

This "feedback loop" is provided by the ideal relationship usually developed between a center and its constituency and is highly significant to the dynamics of a technical field's evolution. Another related important function which the information center can serve is that of a nerve center or nucleus around which concerned users can express their criticism with regard to shortcomings in the literature or other aspects of the communication system in their area, working collectively towards effecting its improvement. As Swanson (1966) has observed: "An information monitoring center with an overview of the total communication process would be in a good position to recognize the emergence of subject areas that for one reason or another stand in need of a review or summary." He goes on to point up their value in galvanizing into action scientists who might otherwise remain individually immobilized. In this sense the specialized information analysis center plays an analogous role to a professional society—in fact, there is often a symbiosis between such groups.*

The role of information-analysis centers in data review and analysis is a highly demanding and significant one. As has been pointed out by Howerton the significant term in "critical data" is *critical* and this function requires creative personnel of the first rank. Doctor Branscomb, in his interesting essay, "Is the Literature Worth Reviewing?" (1968), takes note of the fact that a mere rote extraction of data from the literature will only continue to propagate much that is trivial or erroneous. A critical facility to seriously evaluate the data for consistency and good scientific and experimental practice is necessary to fit the data together in any useful review. The recent establishment of the *National Standard Reference Data System* (NSRDS) within the National Bureau of Standards (and under the aegis of the Federal Council for Science and Technology) is an important step in establishing the resources and organization to organize the nation for this formidable task. (For a discussion of data centers per se as a special case of information—analysis centers, confer Brady and Wallenstein, 1967.)

The opportunity that the perspective of an information-analysis center offers to an individual for an overview of a field and the resultant possibility for creative synthesis and induction is in turn an attraction that draws competent and outstanding technical personnel to these tasks. The

*Cf., for example, the relationships of the professional society of military infrared workers, IRIS, and the community information center in the IR field, IRIA, which complement each other and cooperate in various communication areas for the community of opticists.

presence of such technical expertise together with information and document analysts often aided by requisite resources of modern information-handling technology and mechanization provides a symbiosis which is largely responsible for the success of such centers and which explains their proliferated growth.

However, there are many problems associated with the "care and feeding" of information-analysis centers which should be mentioned. In spite of the possibility for creative work mentioned above, the scientific community has not yet given the approbation to this review and synthesis function which it clearly deserves. PSAC (1963) recognized the importance of this work in a recommendation to the technical community:

> We shall cope with the information explosion, in the long run, only if some scientists and engineers are prepared to commit themselves deeply to the job of sifting, reviewing, and synthesizing information; i.e., to handling information with sophistication and meaning, not merely mechanically. Such scientists must create new science, not just shuffle documents: their activities of reviewing, writing books, criticizing, and synthesizing are as much a part of science as is traditional research. *We urge the technical community to accord such individuals the esteem that matches the importance of their jobs and to reward them well for their efforts.*

One successful technique for providing the technical expertise necessary to this review function is to place the information-analysis center at a location where related research is otherwise supported, thereby taking good advantage of the possibilities for "time sharing" of competent individuals for information-related tasks. Thus the full cost of such personnel need not be borne by the explicit support for the information-analysis center, a not insignificant item in view of the critical financial support problems which such centers normally encounter. This explains the phenomenon that so many information centers function at establishments such as government or university research centers, where continued support for related research is provided. For those organizations who do not enjoy the luxury of continued institutional research support, such as in industrial or even university environments, this problem of maintaining enough related research activity to maintain the essential research expertise related to the information center-functions is a great challenge. In so far as the need arises to compete with other groups for specific research support the related aspects of conflict of interest apply, because of the possibility of unfair advantage accruing to the center by virtue of the proprietary or special advance knowledge which the information base offers.

The problem of information center support is getting more acute as Federal budgets become more competitive, and as these centers proliferate as a result of the pressures of increased specialization. A recent trend has been to try to displace more of the operational costs on to the users who have heretofore been serviced gratis or at a very nominal charge thanks to the overall center support base.

This approach is part of the general trend in the support of information services to shift as much of the costs on to the users as possible. Unfortunately the mechanism for efficient handling of a wide number of user-charges, as well as the receptivity of users for payment of such services, has yet to be proven. The manner in which the market-place offers an opportunity for evaluating the subject services is an obvious positive aspect of this approach to support—especially since an alternative, more sophisticated measure of effectiveness for evaluating information services has been so extremely difficult to identify. However, the using community has unfortunately not yet established a mechanism for properly appreciating the monetary value of *information*, which would prepare it for entering into this type of market-place approach to the subject. In addition, the collective needs of the community would be difficult to anticipate for the purpose of allowing meaningful planning under such a fractionated support structure. In view of the fact that the Federal Government frequently supports the users indirectly through the grant, contract and direct employment route, it may even appear inefficient and somewhat duplicative to pay for all of the overhead and handling charges in the turbulent recirculation of funds within the community, rather than through a direct and singular subsidization of the particular information service. (Provision for user feedback can still be provided through surveys and committee mechanisms.)

Of course, this problem of financing philosophy is not unique to information and analysis centers, and a similar discussion is equally applicable to other elements of the communication system such as journals, technical reports, information and abstracting services, etc. This whole area is searching for an effective *modus vivendi* and must face up to significant issues between the Government, commercial interests and the professional and private-sector communities.

Many of these issues come into clear and explicit focus when the subject of the establishment of a *new* information and analysis center is seriously raised. The most clear-cut case is where a new mission or discipline is explicitly recognized. In recent instances, upon the initiation of new technical legislation or bureau reorganization, the explicit need for an information-analysis center has been recognized and included in the basic authorization or other directive setting up the undertaking. This was the case for the establishment of the National Oceanographic Data Center when the Government re-emphasized the national interest in oceanography. However, subsequent appropriations are usually more difficult when a number of agencies share responsibility, as in this same case. As a new discipline mushrooms into significance there are likely to be a number of concentrations of information significance, and it is not always obvious that a particular institution should be singled out for support as an information-analysis center. Fortunately (or unfortunately, as the case may be) there are not always obvious rewards attendant to the establishment of such centers, so that there is often little competition; when an organization comes forward and suggests itself for identification as the site for such a center, the principal problem is usually that of finding support for such an undertaking.* Commonly the other aspects of information handling precede the establishment of a center—symposia, journals, curricula and centers of excellence in research. As the field reaches a certain critical size the significance of a community center seems to be widely recognized (as in the field of LASERS today, for example). Alternatively, the importance of a subject which is submerged in a wide number of existing activities, seems to take on a sudden significance of its own and seems to warrant explicit attention through an information center format. The field of *sensors* and *remote sensing* is an example of such an area at the moment; one that is also attempting to establish a center and facing the problem of obtaining support.

Some institutions have been able to justify a number of related information and analysis centers by virtue of their symbiotic coexistence within

*This was the case for one of the earliest government-supported community centers; the author recalls his efforts in attempting to get support for the establishment of the *Infrared* Information and Analysis Center (IRIA) at the University of Michigan in 1955.

the same institution. The interrelatedness and efficiency of mutual support by technical personnel and related data bases makes such institutions uniquely qualified as new subjects require center recognition. (Battelle Memorial Institute seems to be the leader in this field, with some dozen or more such centers operating under their cognizance.)

The National Science Foundation (OSIS) has been able to act as midwife at the birth of many information-analysis centers by providing good offices to bring together a wide assortment of using groups and to provide seed money for the establishment of the center until it can be placed on a firmer support base. In these cases, the Foundation *responds* to a felt need on the part of the using community, since it does not wish to be placed in the position of unduly adding to the proliferation of centers by unilateral action.

Clearly the various issues relating to information and analysis centers have enough commonality and are sufficiently related to the other elements of information handling to warrant continued attention. To this end, a number of organizations have arisen to assist in coordination. An *ad hoc* group of operators of such centers have organized several conferences to review their mutual problems, and a specialist group under the aegis of the American Society for Information Science (ASIS) has recently been formed. Also, the above-mentioned COSATI Panel Number Six is rapidly assuming the responsibility of providing a government-wide focal point. NSF and the LC's National Referral Center are interested in *identifying* the various centers and in studying their potentials and problems. However, it would appear that the using community and their private sector professional societies hold in their hands the future of this important aspect of the information handling system.

In connection with the international aspects of scientific communication (cf. Chapter 9), it is interesting to note that the data center-data compilation activity has been a particularly fruitful one for international scientific cooperation. The compilation of the well-known International Critical Tables in the 1920's by a cooperative international group of experts was a noteworthy progenitor to the National Standard Reference Data System. The data centers resulting from the vast accumulation of data obtained during the International Geophysical Year are still functioning cooperatively (cf., for example, World Data Center A in the U.S.), under the principles set forth by the International Council of Scientific

Unions to expedite the compilation of data coming from several hundred cooperating institutions throughout the world.

The mechanism that an international *network* of information-analysis centers might offer, to provide the nucleus around which to construct an international scientific communication system has been raised by Doctor Chalmers Sherwin, who as was previously noted, is especially impressed by the compacting function of such centers so as to make feasible the computer handling and storage of the world's technical information. He visualizes a standardized system of reporting and machine storage which would allow such a network of a few thousand centers to communicate using a common machine language while still permitting decentralized retrieval and local responsiveness to users. (See, for example, his recent presentation to the ICSU–UNESCO information system conference; Sherwin, 1968.) While international systems normally suffer from a surfeit of inertia and a famine of funds, the function of common standards of data and information handling is an essential and relatively inexpensive area in which to make a propitious beginning.

CHAPTER 8

Mechanization and Information Handling

In view of the growing corpus of information and the general frustrations of the community in coping with it, it is understandable that mechanization should be anxiously looked upon for a solution. Nevertheless, in spite of numerous attempts to provide detailed assistance of various degrees of success, via this route, it is fair to say that the problem is much more difficult and sophisticated than most computer enthusiasts have recognized. Some balanced perspective is needed.*

As has been pointed out by the recent "Licklider Panel" (1965) for the President's Science Advisor (on which the author served):

> Some speak as though computers were going, all by themselves, to solve the problems of the information explosion. (Indeed the phrase, "information explosion," tends to occur frequently in the context of computer magic.) Others, perhaps in over-reaction, tend to reject proposed applications of computers without taking time to understand them, and seem to consider it almost fraudulent to mention computers in the context of documents and libraries. . . . It is clear to us that both of the extreme attitudes are wrong. Computers have demonstrated usefulness in applications such as production of *Index Medicus* (The Medlars Project), but they extend no short-term promise of automating all the functions of a library, let alone a national system for scientific and technical communication. For a long time, we shall be dealing with systems that include men as well as machines. Certainly the systems have to be planned and designed by men, and

*It has been interesting to observe this perspective position develop as the post-Sputnik enthusiasm for mechanization and large centralized monoliths gave way to the present conservative eclecticism. Thus see the Congressional hearings and reports by the Humphrey (1962) and Pucinski (1963) Committees.

computers have to be programmed by men. In short, computers offer no magical solutions, but they are potentially such useful tools it would be very wrong to fail to exploit them.

A number of computer constructed indexes have already been mentioned as providing interesting and potentially powerful new approaches, for example, the Science Citation Index and the KWIC or Keyword in Context Index. These are examples of the power of the computer to operate on easily accessible cues in the literature (e.g. citations and title words) to process the data in a new and useful form. However, it should be noted that the intellectual activity, which is the really significant contribution in these instances is preaccomplished in these cases in the author's selection of title terms and in his laborious encounters with pertinent references.

The formulation of key words into an available dictionary or thesaurus is basic to many of these approaches. Starting with the attempts to expand and keep current various organizational schemes for books, such as the Universal Decimal Classification (UDC) and the Library of Congress system, it has been necessary to organize and reorganize as new disciplines develop and new insight into scientific relationships evolve. The struggle for hierarchical organization of the subject matter, while essential and of great significance (see, for example, the recent DoD-wide Thesaurus of Project LEX (1967)), must nevertheless continually fail as new discoveries occur which alter the logical organization of any body of knowledge. One is thus drawn into the more flexible "random access" mode of employing *descriptors*. Retrieval of material can thus be called for on the basis of commonality of descriptors and a search strategy can be devised upon the basis of an intersection of search terms (see Chapter 6, Section A). This is referred to as coordinate indexing, after M. Taube (1963). Manual and various mechnical techniques have been in vogue for some time, employing split-page systems, notched cards and manual sorting needles and the more rapid Hollerith cards and mechanical sorting. A powerful and relatively simple and inexpensive system is the "Peek-a-boo" system which employs a see-through search of area-punched cards to locate the geometrical intersection corresponding to commonality of chosen descriptors.* While these systems function quite satisfactorily for small to

*This system was described by Wildhack and Stern (1958). Commercially available systems of this type are produced by the Jonker Corporation (1960).

medium-size collections of documents, a computer system becomes quite cost-effective for larger collections (say above 5000 documents). Such computer coordinate-indexing techniques have been popular for some time in industry and in large clearinghouse services (e.g. Defense Documentation Center Service) but have suffered somewhat in their requirement for "batch processing" on computers, with the attendant delay to the user and the generally unsatisfactory open-ended time delay relationship. The importance of timeliness of a response to the satisfaction of user needs, while varying from request to request, should not be underestimated. The iterative manner in which scientists and engineers often utilize information, together with the importance of quickly ascertaining whether one is pursuing a relevant and useful literature search, has been one of the reasons for the preference of manual systems (catalogs, library shelf browsing, and files of index cards, for example) over computer responses that offer delay or require translation of the needs of the user through several echelons of respondents. A recent innovation with the advent of so-called "third generation" computers employing the time-sharing mode, together with sophisticated programming which permits real language interrogation via local work environment console access, provides a wonderful new capability allowing the user to carry on a productive dialogue with the computer. It is now feasible in these instances to carry out a direct closed-loop real time search strategy, which may be modified and recycled as the user gets new insight into the literature corpus characteristics.* (For an excellent description of the power of these new techniques for such iterative search procedures see the recent exposition by Licklider and Taylor, 1968.)

A further opportunity of mechanization associated with the above use of computer sorted descriptor information is in the opportunity that large computer *memories* offer for the inclusion of title and abstract information. Using a visual display in conjunction with the console technique described above, one can refer to additional information on a retrieved reference to further check its relevance. Other pragmatic innovations suggest themselves, for example a useful feature tried out in a NASA experiment employing a Bunker-Ramo real-time-shared computer

*Two pioneering efforts in this field of time-shared real time scientific information retrieval systems are those operated by the System Development Corporation (ORBIT) and the MIT Project MAC (and TIPS).

coordinate indexing retrieval system, was the automatic print-out of finally chosen reference citations, selected by the user, on a "slaved" console at the local information center for subsequent hard copy retrieval and forwarding to the user, thus eliminating the tedious and time-consuming chore of filling out and forwarding ordering slips for library processing.

An even more rapid ability to gain access to the selected literature from a machine search is becoming available through the new format of "microfiche" document collections. This inexpensive and easily stored and reproduced microform allows a complete corpus of the documents in an information store, to be maintained in the immediate user console environment so that the material referred to can be searched and copies made of the important sections for retention with no delay. This approach is instrumented in the System Development Corporation's Project "Orbit". A number of techniques are available for rapid access to remote stores of documents in hard copy or microfiche; thus, for example, the exploitation of closed-circuit television interrogation of microfiche files has now been perfected by the Saunders Corporation, and the transmittal of hard copy excerpts has been available for several years, from devices such as the Xerox Corporation's Magnafax Telecopier which transmits over the narrow band-width lines of regular telephone circuits. The importance of this need to facilitate rapid interaction with referenced material as the need arises, as well as the recycling capability that the iterative technique facilitates, should bring a whole new dimension to the scientist and engineers' effective reference to the literature. The barriers of time delay and the consequent breaks in a search and the accompanying intellectual creativity should be greatly reduced and the open-cycle nature of so much of present-day literature searching through intermediaries will be ameliorated.

A related technique will be the powerful searching of citation lists (see Chapter 6, Section A) in mechanized form, which will greatly strengthen the utility of this approach. This innovation occurs in the mechanized Science Citation Index system operated by ISI as well as in the system known as TIPS, under the general MAC-mechanized system at MIT, the latter thus far restricted to the *core journal* literature for Physics.

An experimental innovation, which has been made possible by very rapid search computer operations on stored *full text*, allows a new degree of freedom in performing searches, which is not dependent upon the previous inherent limitations of the assignment of descriptors by editors, or the title terms or even reference citations by authors.* The *full text* of an article can be searched according to any strategy that can be given in Boolean terms to the computer (Bunker-Ramo, General Electric). This is getting amazingly close to the nirvana-like concept of a "reading machine".

A number of other categories of information files are available for mechanized searches. Files on current efforts, or "who is doing research on what subjects", offer new opportunities for expanding and systematizing informal communications. The Science Information Exchange, operated by the Smithsonian Institution, compiles such information on basic research projects, preparing a notice of a research effort for each government or privately sponsored research effort and filing information in computer form on these efforts according to various indexing terms, including the assignment of descriptors by scientific specialists who are responsible for both input and output. A similar and even larger system is being organized within the Government on the basis of information contained in the so-called form 1498, which is the project card providing all pertinent "current effort" information on an internal or external research work unit. It is hoped that the system now in operation between DoD, NASA and AEC can be expanded into a government-wide system, as recommended by a FCST task force under Dr. C. Sherwin (1967). This area of current effort information has been a subject of government concern for some considerable time (cf. the Congressional "Wenk" and "Stern" reports, for the Senate Humphrey Committee (1960), and the House "Elliott" Committee), and it is certainly badly needed to permit greater cooperation and coordination in the wide diversity of research efforts supported by the many government agencies whose missions are often sufficiently blurred in the complexity and dynamics of modern R and D growth.

In the past, most discussions of mechanization have tended to be co-extensive with advocacy of centralization. This debate (see, for example,

*Haibt *et al.* (1967) reported on a study conducted jointly by Time, Inc., and IBM which simulates a full text search system as it might operate within the reference library of a news magazine. They found that a simple full-text search system with user interaction can give retrieval accuracy which is quite competitive with manual search methods.

the Congressman Pucinski, 1963, hearings on a National Center for Documentation and the post-Sputnik discussions of emulating the Soviet VINITI approach) has probably done considerable harm to the advance of mechanization in information handling since the U.S. scientific community has shared the rest of the populace's concern and fright of a "big brother" *centralized* approach to information. In retrospect, much of this concern was probably emotional, but the scientist was especially anxious to preserve his own prerogatives in browsing and searching the literature according to his own work and thought processes, which are fortunately still unprogrammable. The idea of turning all of this over to a central elite of computer experts smacked of the worst offenses conceivable in a "1984 syndrome". (Incidentally, some of this is already visible in connection with the complaints one hears rising up in the Soviet centralized information system as a reaction to VINITI, which has been accused of being "all input and no output".)

We are now beginning to see that the scientist and engineer user, far from being unrelated to the computerized information system, may be coupled to it in a way which was never really possible in the past. Professional societies, authors, users, information and analysis centers and all other organizations which have been playing a role in information handling become more, not less, essential in the mechanized information systems being gradually evolved. Furthermore, the advent of the real-time access techniques described above, offer a whole new dimension to the search and availability factors which *enlarge* the user's ability to interact with the knowledge of his co-workers, cutting across many of the previous barriers of space and time.*

The full strength and power of such systems, if they are to provide the essential features of coherence and commonality but still allow the desired decentralization, will be critically dependent upon the achievement of common standards and other compatability factors, which must be dealt with at national and international levels. Not only language issues and the

*The enhancement of direct scientific communication among distinctly separated colleagues by means of a common computer network link has been graphically demonstrated by Dr. Kirsch (NBS). By "slaving" their consoles they can get a common print-out from a data base, work together in a joint strategy and use the system in lieu of correspondence, etc. They can, furthermore, pool their own data systems for search and communication.

compilation of common thesauri, but the multitude of codens for uniformity of source references in the book, journal and document literature, must be rationally developed and scrupulously adhered to.* Furthermore, the language of machine processing and computer programming must be coordinated, if we are not to substitute a new Tower of Babel which will prevent us from truly enjoying the benefits of this new technology. Fortunately, various people have recognized this growing problem and work has been initiated by various groups moving toward solutions (see, for example, Newman's Information Systems Compatability, 1965). Dr. C. Sherwin has recently reported on his White House Task Force effort to make some progress in this area in the U.S. (1968). He has recommended a specific language, standards and sequencing which should enable communication between the various mechanized systems being developed. Specifically, this representative group has called attention to the facts that: "There is an urgent need for the following common codes: country, language, Serial titles, standard book numbers, organizational entity codes necessary to identify documents and technical reports."

For the continuing responsibility for common standards and mechanization compatability this group deferred to the U.S.A. Standards Institute Committee Z-39, U.S.A. S.I. Committee on Library Work, Documentation, and Related Practices, but suggested that their work be expedited by the provision of a full time secretariat.

There is a related point, which is probably best raised at this time, involving the question of achieving familiarity and effective access to information systems. With the advent of a diversity of information activities

*The need for a thesaurus or concordance of scientific terms is readily apparent when any concept of a network of information system elements is seriously considered. Heald, who was the project leader of the DoD Project LEX thesaurus compilation (q.v.), has pointed out (1967) that effective machine utilization requires considerable discipline: "This is somewhat contrary to the often heard comments that the machine should be the complete slave, developed in such a manner that the long established customs and even tenor of humans can continue unchanged. Such is not the case. Whether admitted or not, to apply mechanized techniques effectively to library type functions and the transfer of information, human beings must make certain changes and perhaps forego certain freedoms with which they have been accustomed." It is possible that the computer itself can make this task less onerous by carrying out a tutorial dialogue with the user in such a way that this discipline is established, as in the iterative display, and dialogue via computer console described above.

and resources, as is becoming increasingly the norm, individuals in the private sector and government are often in a confused state as to where to go for information. Calling on a colleague for assistance (as has been indicated to be the accepted case in industrial-practice surveys) does not really solve the problem but only defers it to another, who unfortunately is very probably equally ignorant of the range of resources. Some groups have proposed ameliorative steps in the nature of central clearinghouses or referral centers to narrow the range to inquiries and improve the important factor of resource "conspicuity". The Library of Congress' National Referral Center for Scientific and Technological Information was set up to be such a conspicuous center and Battelle operates an Information Center on Information (IRC). With the advent of mechanization it now becomes possible to provide the individual with support in the form of general files on *sources of information* which should allow an expansion and updating of awareness of potential sources of information beyond the outdated and often inadequate base referred to in his formal education and training or those that appear from time to time in publications but are soon out of date. This can be among the most powerful information tools supplied by a mechanized system.

A number of additional publications are now available to assist scientific workers with an overview of agency information practices and available systems. The National Science Foundation has funded a continuing survey of "Scientific Information Activities of Federal Agencies" in serial bulletin form (34 to date) to complement their now inactive survey of Federal organization for Scientific Activities (last published in 1962). The Library of Congress' National Referral Center has recently issued a *Directory of Information Resources of the U.S. Federal Government* (June 1967), with a supplement of Government-sponsored Information Resources.

With regard to mechanized systems, the FCST–COSATI Committee has recently published: *Selected Mechanized Scientific and Technical Information Systems*, available from the Government Printing Office, September 1968. NSF has a highly informative continuing publication which includes announcement of new efforts (*Science Information Notes*, bimonthly); their annual compendium on *Non-Conventional, Technical Information Systems in Current Use* reports on techniques still in the research and development stage. Recently the extensive current efforts of the Federal Government in R and D have proven too great for such a compendium procedure and the DDC mechanized current effort file has been expanded and concentrated on this topic under a COSATI task group. Periodic machine print-outs are expected to be available in this format.

The advent of modern and efficient information techniques involving mechanization, microforms and reprography are already revolutionizing the concept of a "library", and imaginative studies of the future of the library by Licklider (1965) and Overhage (1967) portend even more exciting possibilities. The principal dichotomy of library policy; local self-sufficiency vs. sharing of resources (cf. Clapp, 1964) becomes academic when libraries will be interconnected in a network via computer coupling. There will still be the important root problems of intellectual processing of accessions for cataloguing and indexing (see Chapter 6, Section A) and the non-trivial issues surrounding the provision for collection of materials in the first instance. Once these issues are settled and the requisite fair re-imbursement for intellectual property arrived at there promises to be a real quantum jump improvement in the ease with which the intellectual record may be searched and the scholastic use of this record may be accomplished. This should put to an end the long-endured dissatisfactions which have explained why, as V. Bush (1945) observed, "even the modern great library is not generally consulted; it is (only) nibbled at by a few". Thus the delays in accessions and cataloguing (previously referred to), shelving and mis-shelving, binding of journal issues in an annual, inadequate library physical facilities and hours of access, physical removal of the unique copy holdings to satisfy loan requests, frustrating responses to requestors that the item is not on shelf, etc., all should become relics in a museum of horrors. As a pleasant substitute, we can already see the innovations of rapid provision of the full text into "library" repositories on magnetic tape and in microfilm and special library type microfiche, availability in decentralized locations of the full catalogue and indexing information for search (from time-shared computer consoles), rapid reproduction of the material selected from searches as applicable without disturbing the integrity of the original collection for subsequent users, etc.

The modern type of mechanized information retrieval system discussed above, involving a time-shared real-time common language access to a series of data bases, clearly also opens the door to a new expanded type of "current awareness" service which already has been tried in some limited formats as a so-called "selective dissemination of information" service

(SDI). By describing one's so-called information user-profile of *a priori* interests one can be delivered relevant items (in hard copy, microfiche or limited to relevant excerpts only) as they materialize into information banks. Thus, documents which have the right matching of descriptors, key words in titles, citations to key related articles* can be automatically despatched to the subscriber.† Essentially, this technique provides a uniquely assembled "journal" for each subscriber, in composite form "published" to satisfy his explicit interests. There will still be a need for browsing and new unforeseen areas of interest will have to be searched as the occasion arises, but this technique should do much to counter the existing tendencies for the literature to scatter in the multitude of channels where it gets diffused and dispersed. (See, for example, the discussion of this technique in achieving the "journal of the future" (Brown *et al.*, 1967).)‡

It now seems clear that one of the great strengths of computer mechanization of information will be the provision of "hand-tailored" products with speed and great efficiency. However, as discussed earlier in connection with widespread reprography practices, the dangers to the present support base of primary and secondary scientific journals that these systems potentially pose is enormous, and the obviously useful approach of these mechanized techniques must be accompanied by similar inventiveness aimed at achieving a solution to the economic viability of the *source materials*.

Finally, two caveats should be raised at this juncture as we move forward into the new age of "information-affluence". One concerns the fact that any new opportunities for information handling improvements essentially offer "positive feedback", in the language of the engineers, to the communication system. This will have the result of making more

*See, for example, ISI's ASCA service which operates on this basis of common citations and descriptors.

†A large number of SDI services are now available both from government networks (e.g. NASA) and from private organizations; special lists of relevant journals and selected announcements of technical reports are also available. (See, for example, the CAST system operated by the FCSTI.)

‡As predicted by Swanson (1966): "the scientist of the future instead of receiving each month eight journals containing 300 articles of which 6 are of interest, would instead receive several 'repackaged' journals containing 30 articles, of which 20 might be of direct interest and relevance to his specialty. These articles have been culled from perhaps a thousand journals."

efficient the productivity and communication output of scientific workers, with the result that *still more* information is produced more rapidly. Thus the system will expand still further and will only be prevented from instabilities and true "explosion" by the various impedance terms active in the "real world", including limitations on time, money and other resources, personnel, language, thought and assimilation processes, etc.

The second caveat is that we must not lose sight of the human factors involved in the communications system. We must recognize that in the final analysis both the *originator* and the *consumer* of scientific information are themselves members of the technical community. Their limitations on writing and expression skills, and the quality control process on their primary publications, will still provide basic limitations to the data base input. If quality is lost in a mechanized system this only complicates and exacerbates the retrieval process, or as computer experts so aptly characterize this "GIGO" phenomenon: "garbage in—garbage out!"

On the output, the community habits concerning their time and ability to act as a receiver of information, assimilation capabilities and behavior in search and utilization will perforce affect the characteristics of systems to be developed and will determine their real effectiveness.

CHAPTER 9

The International Aspects of Scientific and Technical Communication

It is well recognized that the subject matter of science and technology is truly international; nature is uniform in its behavior and the laws of science obtain not only in all parts of this earth but, as far as is known, throughout the extent of the universe as well.

Furthermore, scientists have long been in the vanguard of international cooperation both in the conduct of science itself and in the sharing of observations and theories with colleagues across geographical boundaries.

Nevertheless, it is essential to realistically recognize that the practitioners of science and technology are almost invariably *nationally* oriented and that their support, with few exceptions, stems from national treasuries with very real and tangible national objectives in mind. The very cogent recognition of science and technology as an important, if not a critical factor, in the political, economic, and national security competitions among nations, places distinct boundary conditions on the modalities and opportunities for international technical communication.

Thus, in spite of the international interdependence credo which seems firmly and happily instilled in the ideology of scientists and engineers, it is fair to state that, at present, the major developments in science and the principal media for publication and communication are *national* in structure. We will now discuss some of the specific issues and phenomena which arise out of this pragmatic *national* nature of science and its communication elements, and then move on to some of the trends toward greater international cooperation.

A. THE LANGUAGE GAP

Unfortunately, the world's scientific community is returning to the Tower of Babel syndrome. It is regrettable, not only from the standpoint of the English-speaking countries, but from the world view as well, that we missed taking advantage of the truly great opportunity which was potentially available after World War II, to take steps to promote English as the "lingua franca" of the scientific and technical world community. Unfortunately, this opportunity now seems to have receded beyond our grasp, as we witness a strong resurgence of the scientific programs of all countries reflect itself through nationalistic aspirations and simple inertia into publication in native languages. Thus, French, Russian, German, Japanese and eventually Chinese are re-establishing important positions in the worldwide distribution of the scientific literature. Statistical analyses show that in some subjects such as geophysics, for example, English is already not the most frequent language of scientific publication, and in many disciplines it has lost its strategic position of containing a majority of the literature population. In almost all areas it is receding from the high watermark that it enjoyed in the 1950s. This is not to say that other countries do not take special pains to be conversant with English language publications, nor should it be ignored that it is still the most widely understood scientific language of world scientists independent of whether they choose to or are free to publish in it.

As Hanson and Phillips (1962) have stated:

> It is therefore alarming to realize that the English scientist, in spite of his favoured position, is at present separated from about 1/3 of a million new texts by language barriers. Moreover, the position is likely to change—for the worse from his point of view.
>
> It seems wise to assume that in the long run the number of significant contributions to scientific knowledge by different countries will be roughly proportional to their populations,* and that except where populations are very small contributions will normally be published in native languages. English scientists must face the probability that in the immediate future an increasing proportion of the world's literature will be in languages which few of them at present can read, notably Russian, Chinese and Japanese.

---

*Or it may be as Price (1967) has indicated that such technical contributions will be proportional to their gross national products which may (hopefully!) increase more rapidly than population and lead to a still greater contribution to the literature scene.

U.S. scientists, facing the increasing need to be aware of the literature stemming from important research in other countries, have a number of approaches available to them, other than the personally onerous solution of learning to read primary materials in other languages.

The support of full, cover-to-cover, translation series of significant foreign journals (primarily Russian) has had a resurgence of popularity. Over seventy such translations series are currently underway, usually with some government subsidy,* although normally under the enlightened aegis of a U.S. professional society with directly related technical interests. Thus, for example, the American Institute of Physics now has some ten translation journals available for subscription and the Institute of Electrical and Electronic Engineers has a comparable number. Although this approach was strongly supported in the "post-Sputnik" period, it soon became evident that it was not the universally indicated economic answer. There followed a shift toward the more selective translation of particular documents that were believed to be of special interest. This selection process was usually based on some review and screening mechanism performed from title-translation lists, or more significantly on the basis of full abstract translations. Two products of the Library of Congress in this area are particularly worthy of mention: *The Monthly Index of Russian Accessions* (MIRA) is a rapidly available translation of titles of journal articles and monographs, and the companion *Current Contents of the Soviet Press*, distributed by the Department of Commerce, is a grouping of more than routine interest journal article abstracts.

In order to establish some bibliographic control over the translations of full texts (articles or books) the Special Library Association and until recently the Department of Commerce have been collaborating on the publication: *Technical Translations*. This monthly publication compiled the titles of translations made by all cooperating private and government groups, together with information on their availability. A quasi-U.S. government group, the so-called Joint Publications Reading Service (JPRS), is responsible for a great many scientific translations, and their

*Some of the impetus for this program, as well as that involving the *ad hoc* translation of technical documents, came from so-called "Public Law 480" funds, which became available for expenditure in other countries, in local currencies, as a result of the U.S. agricultural aid program. Translation services seemed particularly appropriate in many countries as a useful way of employing such funds indigenously.

output is indexed in the Monthly Catalog of the U.S. Government Printing Office. Recently *Technical Translations* became a casualty of the Federal Budget austerity; SLA will continue to publish its own list of privately sponsored translations.

A number of commercial services have provided novel information services in this translation area. Thus, for example, there is a service known as Information Express, which is a discipline-oriented compilation of translated extracts from Russian journal articles which is rapidly circulated to those in related subject areas. The title is taken from a Soviet publication which provides the symmetrical tool for Soviet audiences. The Consultants' Bureau is a technique for consolidating translation services and spreading the cost over the total population of ultimate users. This economic issue is particularly difficult, in that the costs of translation, which are essentially independent of the number of users, can only be properly partitioned over the users by having some *a priori* appreciation of the total demand population—thus some "club" or society grouping of users seems essential. Of course, similarly difficult economic market problems abound in the entire scientific and communication area, as witness the issues surrounding user charges in a number of community areas such as information and analysis centers, or in the journal support questions raised earlier.

The opportunities for amelioration of the language gap that mechanical or computer translation might provide have been anxiously awaited for some time. However, the nuances of language still defy the mechanical robots, and the most recent review of this esoteric subject by a National Academy Committee chaired by Dr. Pierce (1966), of Bell Laboratories, makes it advisable for us not to hold our breath waiting for such a solution. For the next 10 years, at least, it appears that human translators will be required for at least the editing and review of such texts after the computer has made its best attempt—and the full translation by humans might still be more practical and "cost-effective".

A technique that has served the English-speaking community well during the past 20 years has been the attractiveness of U.S. publications to scientists from abroad to publish therein, in English. This practice has undoubtedly been because of the prestige and greater world-wide recognition that these U.S. publications enjoy. A similar phenomenon explains the desire that U.S. scientists have to publish works of great significance initially in the prestige British journal *Nature*. The availability of international scientific journals under the auspices of commercial organizations based in English-speaking countries has also been a very significant force for English language publication of the world's literature. Pergamon Press, and its some 140-odd learned journals which publish predominantly in English, are the front runners in this increasingly competitive area. By maintaining an International Board of Editors for these journals, they apparently surmount the national prestige problem and essentially establish an international *modus operandi* for attracting world-wide contributions.

Over the last 20 years the U.S. has been a very substantial supporter of research by scientists of other nations. Programs by the Department of Defense, the National Science Foundation, etc., have been of great significance in many countries where such support by local sponsors was not as widely available. It was reasonable that the U.S. sponsors should require reporting in English on the results of these grants. The resulting reporting documents were distributed by sponsoring agencies and then accessed by standard documentation centers and made available for secondary distribution and placed under bibliographic control in the same manner as those reports emanating directly from U.S. projects. Due to the rigors of balance of payment economic problems and other budget austerity problems, this type of support is being severely reduced and with it the leverage over language and availability of foreign research.

An evolving related area concerns the manner in which substantial bodies of technical literature might be exchanged with other countries, for mutual benefits. The Federal Council for Science and Technology has recently (March 1968) issued a policy statement to coordinate the programs of the many different agencies active in overseas exchange programs. Included therein among other generally enlightened policy statements encouraging cooperation appears the following: "Agencies of the U.S. Federal Government shall promote international agreement on the use of the most commonly used language for scientific and technical communication. At present, that language is English." Although this wording was probably chosen for its "soft sell" image, it is doubtful that any policy to change that language *from* English would be seriously entertained even if the statistical situation were to be altered!

Further, this policy states that: "Agencies of the U.S. Federal Government shall generally, in exchanging information, seek a reasonable return which may be in the form of publications, information, materials, services or money. . . ."

Thus some leverage may be applied to other countries through this new route of publication exchanges to utilize English, either in translations or to generally provide bibliographic assistance in obtaining local publications for subsequent translations, as a *quid pro quo* in return for U.S. documentation.

In spite of these techniques, there appears to be no question that the increased affluence abroad in developed countries, together with the general realization of the significance of the importance of R and D activities to a nation's continued economic well-being, is going to encourage greater nationalism and pragmatism leading both to further native-language publication and increased emphasis on proprietary and limited access publications, just as it has in the U.S. Thus we see that the problems of international scientific communication are likely to become increasingly exacerbated.

B. INTERNATIONAL COORDINATION OF SCIENTIFIC AND TECHNICAL COMMUNICATION

While it has been the preferred approach for most countries to attempt to "put their own house in order" before attempting international coordination of scientific information, we observe that the time has come where no further delays can be tolerated in trying to formulate more efficient and effective international cooperation. Recognition that something considerably more than *laissez-faire* methods are required have prompted approaches on a number of fronts.

U.S. agencies that act as delegated or dominant agents (as recognized in the FCST Policy Paper) as, for example, the Atomic Energy Commission in the area of Nuclear Physics information have taken the initiative to contact comparable institutions abroad, such as Euratom in nuclear physics, to work out exchange agreements which could assist by providing effective focal points or "nodes" in an international network for the collection and dissemination of relevant materials. Where a multilateral group exists for coordination, this makes this coordination more efficient; however, it is not always possible to identify such intermediary groups and then bilateral agreements are required. Thus in the nuclear physics paradigm, the IAEA (International Atomic Energy Agency) offers the multilateral base for the establishment of an international system and INIS, the International Nuclear Information System, has been spawned with the promise of effective cooperation in this very significant aspect of scientific endeavour. It should be noted that this cooperation stems from more than just altruism on the part of the U.S. It has been observed that already the U.S. produces only 40% of the world's nuclear literature and this percentage contribution is rapidly decreasing as other nations become involved in this exciting research. It is now recognized that the U.S.'s own science and technology draw considerable strength from both an international climate conducive to the exchange of information and a multiplicity of strong scientific and technological centers throughout the world. Of course, even in technical areas where an international mechanism for cooperation exists, the progress is frustratingly slow and the techniques for financial support and operation are cumbersome and often inadequate. (A number of member states and Euratom have lent experts to carry out a detailed study of this project which may permit it to start operating on a small scale in 1970.)

Where a scientific base of cooperation exists cutting across the nations' efforts, and/or an industrial or other constituency base, then inter-governmental cooperation can be expedited. The experience gained in the large cooperative effort of the 1957–8 International Geophysical Year is particularly significant here. A recent observation by one of those most actively involved, Dr. Odishaw (1959), suggests that such international cooperation works best if "there is a truly international problem; governments support and help out but don't control; and scientists themselves organize the work".

Recently a hopeful approach involving some of these recommended ingredients has appeared. Thus the OECD has set up a group to identify the main policy issues on scientific and technical information and the scope for international cooperation in dealing with them. Their findings (OECD, 1968) should give impetus to the urgent further development of means for promoting cooperation and agreement in establishing comprehensive and compatible information systems.

In addition, a pragmatic, yet international, mechanism has begun to gather momentum, based upon a consortium of interests among the UN–

UNESCO and the prestigious if poorly funded International Commission of Scientific Unions (ICSU). A recent conference at UNESCO House (December 1967) brought together the various representatives of government and the scientific community in a businesslike session to plan an attack on this problem. Further meetings are planned and it appears that all the participants now are quite serious about the need for such action. Scientists have, of course, foreseen this need for some time. A good summary of the activities of the international Pugwash group's activities in this scientific communications field was recently given by Bently Glass (1968):

> why is it that in this area, where it should be far easier to achieve real scientific cooperation even less has been accomplished than in the difficult area of agreements to ban nuclear testing and any further spread of nuclear weapons to additional nations? In large part, I suppose, the difficulty traces back to the fundamental structure of scientific organization in most countries. These activities are in the hands of individual scientists and scientific societies, notoriously individualistic and independent. If it has taken several decades to develop the beginnings of national science information services even within biology, chemistry or physics, we should not be discouraged that the international efforts are slow to be born.

While encouraging progress in this interdisciplinary multilateral direction, we should not overlook the progress that can be made in bilateral arrangements (as, for example, in a potential U.S.–U.S.S.R. exchange agreement) or in the potential for advance in particular disciplines or on bibliographic problems through scientific and non-governmental cooperation.* In the latter category, one should mention the progress over the years that has been made possible in cataloging and subject classification standards under the aegis of the Fédération Internationale de Documentation (FID) in The Hague.

An early instance of international cooperation in the exchange of critical data was the National Academy of Science's international undertaking in the preparation of the 1926 International Critical Tables. Hopefully, this character of cooperation can be reintroduced into the present

*In some instances governmental coordination may be too time-consuming and cautious. Thus K. Way (1968) has made a plea for free enterprise in data compilation, urging that committees open a freeway for compilers rather than guiding or channeling their steps. But very often, especially in international areas, the governments can take steps to remove barriers so that these freeways can be used for scientific traffic.

world-wide data efforts now being carried out by the ICSU–CODATA committee,* since as we know, there is only one world to explore and scientific data is universally applicable. (This will be entered into the U.S.'s National Standard Reference Data System.)

Other phases of international coordination, which are needed in the interests of more effective and efficient information transfer, concern the areas of *current efforts* and the identification of related scientific workers and their investigations and resources, and the area of scientific conferences† and informal communication. Two U.S. activities which undertake to cover this to some degree are the Library of Congress' *World List of International Meetings*,‡ and the Smithsonian Institution's Scientific Information Exchange. Both institutions play a significant *non-governmental role* in the international communication area.

With regard to mechanization, in addition to the formalization of standards for compatibility of the various elements of an international network, the evolving network will require communication improvements. Here the great potential of communications satellites has already been noted§ and promises to provide a major tool for expediting world-wide scientific coherence.

In conclusion, we believe that on balance the possibilities for international cooperation in information handling are significantly improving. While there will undoubtedly be a keen competition to exploit science and technology for national proprietary objectives and while we unfortunately see no immediate reduction in the removal of the secrecy constraints on military related research and development publications, there is no question that the positive aspects of international cooperation and interdependence of science and technology are gaining widespread recognition and approbation. The keen concern for a conservation of scarce

*See the description of this effort given by H. Brown (1967).

†There has recently been a greater scrutiny of attendance at such meetings (Noyes, 1968), and also some analysis of their information functions (Compton and Garvey, 1967).

‡A number of other privately supported announcement bulletins are available, cf., for example, *World Meetings*, and *Technical Meetings*.

§See, for example, various UNESCO studies and papers such as that of Campbell, 1966. In the U.S., President Johnson has appointed a Task Force (E. Rostow, Chairman) to examine the ways in which INTELSAT technology can be applied to the related education and scientific communication opportunities.

intellectual resources and the desire of mankind to solve its environmental race against time demand urgent and high quality attention to the national and international good practices of information use and communication.

APPENDIX A

Taxonomy of the Technical Report Literature

THE report literature spans the spectrum from the "interim" type format of the contract periodic progress report (monthly or quarterly, as the case may be) on up the formal hierarchy to the monograph or book equivalent. This outline, while not exhaustive, describes the general archtypes in the phylum together with some discussion of the review process and availability of each type, etc.

1. *The individual author's "preprint"*

Before a technical work is published anywhere, it is customary to process the manuscript for circulation among colleagues for review, etc. Often these "preprint" materials wind up with a somewhat more formal designation (such as RAND Corporation Papers "P's") and are cited or indexed. While editorial and drafting type assistance is often provided by the author's institution, there is a very wide variation in the amount of institutional review of such material. The RAND Corporation P's, for example, carry the notice that the material is merely reproduced as a service to the staff, but does not represent a corporate study. Many university departments exercise no review of material intended for submission to journals, since they assume that the arrangement is one purely between the author and the journal, which medium is expected to exercise its own review process.

There is rarely an explicit indication on "preprints" specifying the journal where the work is to be published, since at this stage the author is probably not certain; in fact, the journal article may be significantly modified when eventually published.

Another type of "preprint" report is the speech or paper to be given at a symposium, the proceedings of which may appear in collected form in a number of different ways (discussed below).

The significant advantage of the preprint report is its early existence, while the work is still in a nascent condition. Distribution is often made to a primary group of colleagues or members of the author's "invisible college". More formal and wider dissemination of such preprint literature, through so-called information exchange groups, has been criticized as a proliferation of unreviewed materials by journal editors and others who feel that the speedy letters-type journals are the preferred dissemination media (see p. 71).

Inevitably preprint reports find their way into clearinghouses, depositories and information centers, where they are indexed, abstracted and announced and often reproduced for secondary dissemination.

2. *The corporate "proposal-type" report*

While this format occupies the time and efforts of some of the best minds in America, the fact that this type of material is usually not circulated beyond the staff of the prospective customer, and often is considered of a proprietary nature, indicates that we should give it only passing notice.

It is interesting to note that even here, there is reason to believe that the informal communication channels have been at work much earlier in conveying the details of the proposal from industry to government evaluators and that in many cases the submission of the proposal is a pro forma activity.* Nevertheless, for the Government's own purposes, however, there is probably a need for improved handling of this format for effective

*See, for example, Roberts (1968) who notes:

"As factual evidence of this personal information flow process, the winners of R & D contract awards differ from the losers in various ways:

The winners had done more prior work with the government agency involved;

They were better acquainted with the government technical initiator;

They tended to employ fewer technical writers (and fewer consultants) to beef up the quality of their written proposals."

evaluation and comparison with related on-going efforts and for resource correlation. (Through the SIE and other current effort systems.)*

3. *Institutional reports*

Organizations and their various sub-elements feel the need, for one reason or another, to issue periodic reports of their activities and progress. Thus we have the series of annual reports of government agencies, foundations, corporations, societies, laboratories, etc. While much of this material is directed at lay public audiences for purposes of budget justification and image enhancement, it is fair to state that knowledge of programs, resources, progress and plans of a technical nature are often best stated coherently in these formats. For example, the report of the President to Congress on Space, and the Space Science Board's Report to the international COSPAR convention, are believed to be among the most comprehensive surveys of space technology and related science programs extant. Publication listings and bibliographies are also considered valuable aspects of these compendia.

In view of the great loss by scattering and diffusion of technical contributions into the world's communication channels in the normal course of specialized technical communication, the synthesis and overview that these organizational reports provide will probably continue to justify their existence.

4. *The contract "progress report"*

This is believed to be the most populous species of technical report in circulation. While primarily directed at the "sponsor" to provide him with a periodic assessment of the problems and progress under the contractual effort, it is often given primary distribution to the sometimes extensive group of interested workers in related problem areas so that rapid interaction, and perhaps utilization, of findings can occur. Since the number of contracts (and sub-contracts, which present more of a problem as far as access is concerned since their distribution is often controlled by the prime contractor) by various government agencies is probably of the

*A recent DoD Conference dealing with circulation and clearinghouse functions for proposals recommended increased formal attention to this type of report.

order of several hundreds of thousands a year, and assuming only quarterly reporting (many contracts call for monthly reporting), we see that the number of individual technical progress report "titles" in circulation or in archives probably numbers in the millions.

While it is true that many of these periodic progress reports are overtaken and refined in the Final Report issued after the effort is completed, it is also true that many are not (that is they are referenced but not repeated in the final report).

Clearinghouses do apply some filtering of non-substantive progress reports before secondary announcement or abstracting. However, since one man's noise is another man's signal, it is probably not wise to have that filtering done by unskilled personnel. For that reason, we believe that the specialized information and analysis centers offer the best vantage point for reviews, indexes, abstracts and quality rating of these materials for subsequent retrieval and advisory functions.

5. *The "Final Report" on a technical contract effort*

This species is probably the most valuable specimen in the collection. It is written from a vantage point of overview of technical accomplishments of the effort and usually has the benefit of considerable editorial support and review.

Nevertheless, there are wide variations in the quality and scope of these reports, as there must be in any area involving so many different institutions both on the writing and contracting ends of the transaction. Furthermore, the formats, serialization, abstracting, indexing, distribution and other aspects of the report handling also differ widely.

Steps are being taken in the Government, however, to apply some uniformity (hopefully uniformly high) of standards to report format and indexing. (Cf., for example, the DoD form 1473 appendage to reports giving report vital statistics which COSATI's subcommittee on Report Format is considering as a uniform standard.)

6. *The "separate", topical, technical report*

In addition to progress reports and final reports, contract efforts normally publish special separate topical reports which come closest to journal articles in style and type. They may originate at the instigation of the sponsor or flow from the same technical desire of technical staff members to be heard when they have something to say, as in the case of journal articles.

Called alternatively, research memoranda, research notes or documents, technical memoranda, etc., they are often subsequently submitted to journals where they are usually published in abbreviated forms. Some institutions later apply corporate or government laboratory covers to the journal article and issue the resultant hybrid as a technical report, further confusing the issue, for readers, clearinghouses and indexing and abstracting services (with the somewhat doubtful justification that the journals may not be uniformly accessible).

It would appear to us that this type of report is most easily the legitimate target of journal editors who would rather prefer having the worthwhile material in these reports published in journals (after refinement and screening by the proper expert review, etc.) and the non-worthwhile materials not issued at all. As discussed earlier, there is much to recommend this point of view for all concerned, not the least of whom is the technical author.

7. *The "book" in report form*

The need for reviews, survey and state-of-the-art reports is well known. Specialized information and analysis centers, special author groups commissioned by agency contracts,* not-for-profit analytical organizations, and to an increasing extent industrial laboratories, now generate such survey-type materials in large numbers.†

Even those items destined to be commercially published often first appear in report form at a much earlier date. While these items probably receive the most critical editorial and technical review of any items in the reports series, the ultimate publication in book form has the advantage of more time for refinement. However, in view of the earlier availability of

*e.g. the AEC often advertises for qualified candidates to prepare these materials under individual and institutional grant support; NASA has a comparable program.

†Many organizations use special series designations for such reports, such as RAND Corporation R's, to distinguish them as a quality product.

the information and reduced cost to the individual in obtaining the report, one can easily see why there would be desires to maintain this format.

8. *Committee-type reports*

The multitude of scientific advisory committees that have arisen in various aspects of our government and private sectors reflect our plural society, and its consensus-seeking behavior. These prestigious groups normally present their conclusions and findings together with their studies and bibliographic annotations in report format; thus we have reports by the President's Science Advisory Committee on a multitude of science policy subjects, Congressional advisory staff reports, hearings, etc., National Academy of Science—National Research Council reports of all types, reports of international scientific bodies, etc.

Report series descriptions, dissemination and availability, indexing and abstracting, retrieval, etc., follow widely varied practices and unfortunately often leave a great deal to be desired.

References

ABELSON, P. (1966) *Science* **154**, 727.

AUERBACH CORP. (1965) DoD user-needs study, *Phase 1*, vols. **1, 2** (AD 615501,2).

BARBER, B. and HIRSCH, W. (Eds) (1962) *The Sociology of Science*, The Free Press, New York.

BATTELLE MEMORIAL INSTITUTE (1967) *Industrial Utilization of the National Information Resource for Science and Technology*, Columbus.

BARR, K. P. (1967) Estimates of the number of currently available scientific and technical periodicals, *J. of Doc.* **23**, 110–116.

DE B. BEAVER, D. (1964) A statistical study of scientific and technical journals, Yale Univ. N.S.F. GN-299.

BOUTRY, G. A. (1959) The ICSU Abstracting Board, its story, ideas, methods and aims, *ICSU Review* **1**, 113–137.

BOUTRY, G. A. (1959a) *ICSU Abstracting Board: To Improve the Distribution of Scientific Information*, Paris.

BRADY, E. L. and WALLENSTEIN, M. B. (1967) *Science* **156**, 754.

BRANSCOMB, L. M. (1968) Is the literature worth reviewing? *Scientific Research*, **3**, 49–57.

BROOKS, H. (1967) Applied science and technological progress, *Science* **156**, 1706–1712.

BROWN, H. (1967) *Science*, **146**, 751.

BROWN, W. S., PIERCE, J. R. and TRAUB, J. F. (1967) The future of scientific journals, *Science* **158**, 1153.

BUSH, V. (1945) As we may think, *Atlantic Monthly* **176**, 101–108.

BUSH, V. (1967) *Memex Revisited, in his Science is not Enough*, William Morrow & Co., New York.

CAMPBELL, H. C. (1966) Some implications for libraries of communications satellites, *Unesco Bull. for Libraries* **XX**, 129–133, 139.

CLAPP, V. W. (1964) *The Future of The Research Library*, Univ. of Illinois Press, Urbana.

CLAPP, V. W. (1968) *Copyright—A Librarian's View*, Association of Research Libraries, New York.

COBLANS, H. (1965) The communication of information, in Goldsmith, M., *Society and Science*, Simon & Schuster, New York.

COBLENTZ, W. W. (1951) *From the Life of a Researcher*, Philosophical Library, New York.

COMMONER, B. (1965) The integrity of science, *Amer. Scientist* **53**.

COMMONER, B. (1966) *Science and Survival*, The Viking Press, New York.

COMMITTEE ON SCIENTIFIC AND TECHNICAL INFORMATION (COSATI), (1964) *COSATI Subject List*, Federal Council for Science and Technology, AD 612200.

COMMITTEE ON SCIENTIFIC AND TECHNICAL INFORMATION (COSATI), (1965) *Recommendations for national document handling systems in science and technology*, Federal Clearinghouse PB 168267.

COMMITTEE ON SCIENTIFIC AND TECHNICAL INFORMATION (1967) *Task Force on Copyright Problems*, Hinton, H., Ch.

COMMITTEE ON SCIENTIFIC AND TECHNICAL INFORMATION (COSATI) (1968) *Panel for a Standard Format for Scientific and Technical Reports Prepared by Contractors or Grantees*, Grewell, W., Ch.

COMMITTEE ON SCIENTIFIC AND TECHNICAL INFORMATION (COSATI) (1968a) *Proceedings of the Forum of Federally Supported Information Analysis Centers*, Federal Clearinghouse.

COMMITTEE ON SCIENTIFIC AND TECHNICAL INFORMATION (COSATI) (1968b) *Directory of Federally Supported Information Analysis Centers*, Federal Clearinghouse.

COMPTON, B. E. and GARVEY, W. D. (1967) Information functions of an international meeting, *Science* **155**, 1648–1650.

CONRAD, C. C. (1967) Coordination and integration of technical information services, *J. of Chem. Doc.* **7**, 111–115.

"CRAWFORD TASK FORCE" (G. Abdian, J. Crawford, W. Fazar, S. Passman, R. Stegmaier, J. Stern) (1962) *Scientific and Technological Communication in the Government*, Office of Special Assistant to the President for Science and Technology, AD 299545.

CRISMAN, T. L. (1967) In-House Publications—can they endanger rights in technical information? *J. of the Patent Office Society* **XLIX**, 549–565.

CUADRA C. (Ed.) (1967) *Annual Review of Information and Technology*, vol. 2, Interscience, New York.

DARBY, R. L. and VEAZIE, W. H. (1968) Writing a state-of-the-art report, *Mats. Res. and Standards*. **8**, 28–32.

DEPARTMENT OF DEFENSE (1964) *Glossary of Information Handling*, DDC.

DEPARTMENT OF DEFENSE (1967) *Thesaurus of Engineering and Scientific Terms*, Project LEX, DDC AD 672000.

DEPARTMENT OF DEFENSE (1968) DDC explains policy changes, *Defense Industry Bull.* pp. 30–32.

ELIAS, A. W. (1967) *Third Conference on Technical Information Center Administration*, Spartan Books, New York.

ELLIOTT, C., Ch. (1964) *Report of the Select Committee of the House of Representatives on Government Research*, Study No. 4, *Documentation and Dissemination of Research and Development Results*, GPO.

FEDERAL COUNCIL FOR SCIENCE AND TECHNOLOGY (1968) *Policies Governing the Foreign Dissemination of Scientific and Technical Information by Agencies of the U.S. Government*, Office of Science and Technology.

FID (1962) The content, influence and value of scientific conference papers and proceedings, *Unesco Bull. for Libraries* **16,** Nos. 3 and 4 (in two parts).

FORNEY, G. D., Jr. (1963) *Encouraging Review Articles in the Sciences*, Office of Science and Technology.

FRECHETTE, V. D. and CONDIT, R. H. (1967) *Physics Today* **20,** 53.

FRY, G. and Associates (1962) *Survey of Copyrighted Material Reproduction Practices* in *Scientific and Technical Field*, Chicago, National Science Foundation.

FUCCILLO, D. A., Jr. (Ed.) (1967) *The Technical Report in the Biomedical Literature*, A Workshop at the National Institutes of Health, Bethesda.

GARVEY, W. D. and GRIFFITH, B. C. (1964) Scientific information exchange in psychology, *Science* **146,** 1655–1659.

GARVEY, W. D. and GRIFFITH, B. C. (1965) *Reports of the American Psychological Association's Project on Scientific Information Exchange in Psychology*, vol. 2 (NSF).

GARVEY, W. D. and GRIFFITH, B. C. (1967) Scientific communication as a social system, *Science* **157,** 1011–1016.

GARVEY, W . D. and GRIFFITH, B. C. (1967a) *Professional Societies and Information Exchange*, A series of reports by the Johns Hopkins Univ. Center for Research in Scientific Communication, for NSF.

GARWIN, R. L. (1968) Strengthening military technology, *Sci. and Tech.* **82,** 22–27.

GINZBERG, E., HIESTAND, D. L. and REUBENS, B. G. (1965) *The Pluralistic Economy*, McGraw-Hill, New York.

GIPE, G. A. (1967) *Nearer to the Dust: Copyright and the Machime*, the Williams and Wilkins Co., Baltimore.

GLASS, B. (1968) Pugwash interest in communications, *Science* **159,** 1328–1331.

GOLDHOR, H. (1966) *Proc. of the 1966 Clinic on Library Applications of Data Processing*, Univ. of Illinois.

GOOD, R. H. (1967) The big trouble with scientific writing, *Science* **158,** 1375.

GOTTCHALK, C. M. and DESMOND, W. F. (1963) Worldwide census of scientific and technical serials, *Amer. Doc.* **14,** 188–194.

GOUDSMIT, S. A. (1966) Is the literature worth retrieving? *Physics Today* **19,** 52–55.

GRAHAM, W. R. *et al.* (1967) *Exploration of Oral/Informal Technical Communications Behavior*, American Institutes for Research Final Report AD 669586.

GRAY, D. E. and ROSENBORG, S. (1957) Do technical reports become published papers? *Physics Today* **10,** 18.

GRAY, D. E. (1962) Information and Research—blood relatives or in-laws? *Science* **137,** 263–266.

GREEN, D. (1967) Death of an experiment, *Int'l Sci. and Tech.*, May, pp. 82–88.

GREENBERG, D. S. (1968) *The Politics of Pure Science*, New American Library, New York.

HAGSTROM, W. O. (1965) *The Scientific Community*, Basic Books New York.

HAIBT, L. *et al.* (1967) *Retrieving 4,000 References without Indexing, in Information Retrieval—The User's Viewpoint—An Aid to Design*, A. B. TONIK (Ed.), Int'l Information Inc.

HANSON, C. W. and PHILLIPS, M. (1962) *The Foreign Language Barrier in Science and Technology, Aslib, London.*

HASHOVSKY, A. G. and DOWNIE, C. S. (1967) *Selective Dissemination of Information in Practice: Survey of Operational and Experimental SDI Systems*, Office of Aerospace Research report 67–012.

HEALD, J. H. (1967) *The Making of TEST (Thesaurus of Engineering and Scientific Terms)*, Office of Naval Research AD 661001.

HEILPRIN, L. B. (1967) Technology and the future of the copyright principle, *Phi Deltan Kappan*, Jan., 220–225.

HEILPRIN, L. B. (1967a) *Effects of Copyright and of Journal Economics on Communication in Science and Education*, American Society for Engineering Education, East Lansing, Michigan.

HERNER, S. (1954) Information-gathering habits of workers in pure and applied science. *Ind. and Eng. Chem.* **46,** 228–236.

HERNER, M. and HERNER, S. (1959) The current status of the government research report in the U.S.A., *Unesco Bull. Lib.* **13,** 187–196.

HERNER AND Co. (1961) *Factors Governing the Announcement and Publication of U.S. Government Research Reports*, NSF—C153.

HERRING, C. (1968) Distill or drown: the need for reviews, *Physics Today* **21,** 27–33.

HUMPHREY, H. H. Sen., Ch. (1961) *Coordination of Information on Current Scientific Resarch and Development Supported by the U.S.*, Report No. 263, 87th Congress (U.S. Senate Committee on government operations).

ICSU ABSTRACTING BOARD (1967) *Some Characteristics of Primary Periodicals in the Domain of Physical Sciences.*

The Information Deluge (1967) *The Johns Hopkins Magazine*, Nov. 1967, 2–34.

JONKER, F. (1960) The Termatrex inverted punched card systems, *Amer. Doc.* **11,** 305–315.

JUHASZ, S. and AMMINGER, O. (1965) *Critical Abstracts*, paper presented at the Seventh Annual Institute of Information Storage and Retrieval, American Univ.

KASH, D. E. (1967) The General Pattern of Space Cooperation Information Exchange, Chapter 5 in *The Politics of Space Cooperation*, Purdue Univ. Studies.

KENT, A. (1965) *Specialized Information Centers*, Spartan, New York.

KENT, A. (1966) *Textbook on Mechanized Information Retrieval*, 2nd ed., Interscience, New York.

KOCH, H. W. (1968) A national information system for physics, *Physics Today* **21,** 41–49.

KOCHEN, M (Ed.) (1967) *The Growth of Knowledge: Readings on Organization and Retrieval of Information*, John Wiley, New York.

KOCHEN, M. (1967a) Book-indexes as building blocks for a cumulative index, *Amer. Doc.* **18,** 59–66.

KRONICK, D. A. (1962) *A History of Scientific Periodicals—The Origins and Development of the Scientific and Technological Press*, Scarecrow Press, New York.

KUNEY, J. H. (1963) Economics of journal publication, *Amer. Doc.* **14,** 238–240.

LIBRARY OF CONGRESS (1965) *A Directory of Information Resources in the United States in Physical Sciences, Biological Sciences and in Engineering*, GPO.

LICKLIDER, J. C. R. *et al.* (1965) *Report of the Panel on Scientific and Technical Communications*, Office of Science and Technology.

LICKLIDER, J. C. R. *et al.* (1965a) *Libraries of the Future*, The MIT Press, Cambridge.

LICKLIDER, J. C. R. *et al.* (1966) A crux in scientific and technical communication, *Amer. Psych.* **21,** 1044–1051.

LICKLIDER, J. C. R. *et al.* (1968) The computer as a communication device, *Sci. and Tech.*, April, pp. 14–20.

LOOSJES, TH. P. (1967) *On Documentation of Scientific Literature*, Butterworths, London.

MACHLUP, F. (1962) *The Production and Distribution of Knowledge in the United States*, Princeton Univ. Press, Princeton.

MADDOX, J. (1967) Is the literature dead or alive? *Nature* **214,** 1077–1079.

MARQUIS, D. G. and ALLEN, T. J. (1966) Communication patterns in applied technology, *Amer. Psych.* **21,** 1052–1060.

MARKE, J. J. (1967) *Copyright and Intellectual Property*, The Fund for the Advancement of Education.

MEADOW, C. T. (1967) *The Analysis of Information Systems: A Programmers Introduction to Information Retrieval*, John Wiley, New York.

MENZEL, H. (1966) Scientific communication: five themes from social science research, *Amer. Psych.* **21,** 999–1004.

MERTON, R. K. (1962) Priorities in scientific discovery, Chapter in Barber's *The Sociology of Science*, The Free Press, Glencoe, L.I.

MERTON, R. K. (1965) The Ambivalence of Scientists, in *Science and Society*, Ed. by Kaplan, N., Rand McNally & Co., New York (originally published in the *Johns Hopkins Univ. Hospital Bulletin*, 1963).

MERTON, R. K. (1968) The Mathew Principle, *Science* **159,** 1056–1063.

MILLER, E. E. (1952) The genesis and characteristics of report literature, *Amer. Doc.* **3,** 91–94.

MORAVCSIK, M. J. (1966) Pro-physics information exchange—a communication experiment (a debate on pre-print exchange), *Physics Today* **19,** 62 ff. and 71.

MORAVCSIK, M. J. (1968) The status-discussion meeting as an antidote to super conferences, *Physics Today* **21,** 48–49.

NATIONAL ACADEMY OF SCIENCES (1966) *Report of the Materials Advisory Board's Ad Hoc Committee on Principles of Research-Engineering Interaction*, Pub. No. MAB-222-M.

NATIONAL FEDERATION OF SCIENCE ABSTRACTING AND INDEXING SERVICES (NFSAIS) (1963) *A Guide to the World's Abstracting and Indexing Services in Science and Technology*, Report No. 102.

NATIONAL LIBRARY OF MEDICINE (1963) *The Medlars Story (Medical Literature Analysis and Retrieval System)*, Dept. of H.E.W.

NATIONAL SCIENCE FOUNDATION (1958) *Directory of Specialized Information Analysis Centers* (Battelle).

NATIONAL SCIENCE FOUNDATION (1964) *Characteristics of Scientific Journals* (1949–1959), Office of Science Information Service N.S.F. 64–20.

NEAL, J. P. (1967) Why bury the knowledge? FID Conference, Tokyo (33rd).

NEWMAN, S. M. Ed. (1965) *Information Systems Compatibility*, Spartan Books, New York.

NORTH AMERICAN AVIATION (Autonetics Division) (1966) *DoD User-Needs Study*, Final Report, Phase 2: *Flow of Scientific and Technical Information Within the Defence Industry*.

NOYES, W. A. (1968) *International Scientific Meetings*, National Academy of Sciences.

ODISHAW, H. (1959) *International Cooperation*, Int'l Sci. and Tech., Prototype.

OECD (1968) A scientific and technical information policy, *The OECD Observer* **33**, 36–38.

OVERHAGE, C. F. J. (1967) Science libraries: prospects and problems, *Science* **155**, 802–806.

OVERHAGE C. F. J. (1967a) *Project Intrex*, MIT Press.

PASSMAN, S. (1963) Utilization of scientific and technological information, *Proc. of the 1st U.S.A.F. Scientific and Technical Information Conf.* AD 450 000, 51–56.

PASSMAN, S. (1968) *Task Group Report on The Role of The Technical Report in Scientific and Technological Communication*, COSATI, Federal Clearinghouse. PB 8180944.

PASTERNAK, S. (1966) Is journal publication obsolescent? *Physics Today* **19**, 38–43.

PASTERNAK, S. (1966a) Criticism of the proposed physics information exchange, *Physics Today* **19**, 63.

PENG, K. C. (1967) *The Design and Analysis of Scientific Experiments*, Addison-Wesley Publishing Co., Chicago.

PEREZ-VITORIA, A. (1963) A code of good practice for scientific publications, *Amer. Doc.* **14**, 241–244.

PHELPS, R. H. and HERLING, J. P. (1960) Alternatives to the scientific periodical, *Unesco Bull. for Librs.* **14**, 69–71.

PIERCE, J. R. (1966) *Language and Machines: Computers in Translation and Linguistics*, National Academy of Sciences Pub. No. 1416.

PRESIDENT OF THE U.S. COMMISSION ON THE PATENT SYSTEM (1966) *To Promote the Progress of Useful Arts, in an Age of Exploding Technology*, GPO.

PRESIDENT'S SCIENCE ADVISORY COMMITTEE (PSAC) (1963) *Science, Government and Information*, A. Weinberg, Ch. GPO.

PRICE, DEREK J. DE SOLLA (1961) *Science Since Babylon*, Yale Univ. Press, New Haven.

PRICE, DEREK J. DE SOLLA (1963) *Little Science, Big Science*, Columbia Univ. Press, New York.

PRICE, DEREK J. DE SOLLA (1964) Ethics of scientific publication, *Science* **144**, 655–657.

PRICE, DEREK J. DE SOLLA (1965) The scientific foundations of science policy, *Nature* **206**, 233–238.

PRICE, DEREK J. DE SOLLA (1965a) Is technology historically independent of science? A study in statistical historiography, *Technology and Culture* **6**, 553–568.

PRICE, DEREK J. DE SOLLA (1965b) The Science of Science, Chapter 14 in Goldsmith's *Society and Science*, Simon & Schuster, New York.

PRICE, DEREK J. DE SOLLA (1967) Research on Research, Chapter 1 in D.L. Arm's *Journeys in Science (Small steps—Great Strides)*, Univ. of New Mexico Press, Albuquerque.

PRICE, DEREK J. DE SOLLA (1967a) Nations can publish or perish, *Sci. and Tech.*, October 1967, p. 84.

PRICE W. J. (1967) The key role of a mission-oriented agency's scientific research activities, *Proc. of the Symposium on Interaction of Science and Technology*, Univ. of Illinois Press, Urbana.

PUCINSKI, R. C., Congressman, Ch. (1963) *Hearings and Report of the Ad Hoc Subcommittee on a National Research Data Processing and Information Retrieval Center*, vols. 1 and 2, GPO.

REIF, F. (1961) The competitive world of the pure scientist, *Science* **134**, 1957–1962.

RONCO, P. G. *et al.* (1965) *Characteristics of Technical Reports that Affect Reader Behavior: A Review of the Literature*, Institute for Psychological Research, Tufts Univ, PB 169409 Federal Clearinghouse.

ROBERTS, E. B. (1968) The myths of R and D management, *Sci. and Tech.* **80**, 40–46.

ROSENBERG, V. (1967) Factors affecting the preferences of industrial personnel for information gathering methods, *Infor. Stor. and Retriev.* **3**, 119–127.

ROSENBLOOM, R. S. and WOLEK, F. W. (1967) *Technology, Information and Organization: Information Transfer in Industrial R and D.*, Harvard Univ. School of Bus., 2 vols. mimeo.

RUBINOFF, M. Ed. (1965) *Toward a National Information System*, Spartan Books, Washington D.C.

RUEBHAUSEN, O. M. (1967) Foreword in Westin, A.F., *Privacy and Freedom*, Atheneum Pubs., New York.

SCOTT, A. F. (Ed.) *Survey of Progress in Chemistry*, vol. 3, Academic Press, New York, Preface, p. vii.

SHERWIN, C. (1967) *Task Force Report on a Governmentwide Current Effort Retrieval System, Federal Council for Science and Technology*, Federal Clearinghouse.

SHERWIN, C. (1968) *Report of the Task Group for the Interchange of Science and Technology Information in Machine Language (ISTIM)*, Office of Science and Technology, GPO.

SIMPSON, G. S. and MURDOCK, J. W. (1967) Dollars and secrets, *Amer. Doc.* **18**, 110–111.

SOPHAR, G. J. (1967) Committee to Investigate Copyright Problems Affecting Communication in Science and Education (CICP), paper delivered to the Council of Biology Editors, May 8, 1967.

SOPHAR, G. J. (1967a) *Final Report on the Determination of Legal Facts and Economic Guideposts with Respect to the Dissemination of Scientific and Educational Information as it is Affected by Copyright*, Federal Clearinghouse PB 178463.

SPECIAL LIBRARIES ASSOCIATION (SLA) (1962) *Dictionary of Report Series Codes.*

SPECIAL LIBRARIES ASSOCIATION (SLA) (1967) *Proceedings of the Regional Workshop on the Report Literature* (Albuquerque, Nov. 1965), Western Periodicals Co., North Hollywood, Calif.

SPROULL, R. L. (1967) Paper at the AAAS Dec. 1967 N.Y. Meeting, Panel on Science and Secrecy, Is Secrecy in Science Ever Justified?

SWANSON, D. R. (1966) Scientific journals and information services of the future, *Amer. Psych* **21,** 1005–1010.

SWANSON, D.R. (1966a) On improving communication among scientists, *Bull. of the Atom Sci.* **22,** 8–12.

SYSTEM DEVELOPMENT CORP. (1966) *A System Study of Abstracting and Indexing in the United States* (COSATI), Federal Clearinghouse PB 174249.

SYSTEM DEVELOPMENT CORP. (1967) Document storage and retrieval, special issue of the *SDC Magazine*, September 1967, Santa Monica.

SYSTEM DEVELOPMENT CORP. (1967a) *National Document-Handling Systems for Science and Technology*, John Wiley, New York.

TAUBE, M. (Ed.) (1953) *Studies in Coordinate Indexing*, Documentation Inc., Bethesda.

THOMPSON, J. I. *et al.* (1962) *Study of the Federal Government's System for Distributing its Unclassified R and D Reports*, NSF C-224.

THOMPSON, M. S. (1962) Peek-a-boo index for a broad-subject collection, *Amer. Doc.* **13,** 187–196.

THORPE, W. V. (1967) International statement on information exchange groups, *Science* **155,** 1195–1196.

URQUHART, D. J. (1962) The National Lending Library for Science and Technology, Boston Spa, Yorkshire, U.K., Brochure, Department of Scientific and Industrial Research, HMSO, U.K.

VOIGHT, M. J. (1961) *Scientists' Approaches to Information*, Amer. Lib. Assoc. ACRL Monograph No. 24 (LC-61-10543).

WALL, E. (1967) A rationale for attacking information problems, *Amer. Doc.* **18,** 97–103.

WATSON, J. D. (1968) *The Double Helix*, Atheneum Pubs., New York.

WAY, K. (1962) Waiting for Mr. Know it All, *Physics Today* **15,** 22–29.

WAY, K. (1968) Free enterprise in data compilation, *Science* **159,** 280–282.

WEAVER, W (1967) *Science and Imagination*, Basic Books, Inc., New York.

WEIL, B. H. (1954) *The Technical Report, Its Preparation, Processing, and Use in Industry and Government*, Reinhold Pub. Co., New York.

WEINBERG, A. (1963) *Science, Government, and Information*, President's Advisory Committee's Report, GPO.

WEINBERG, A. (1967) *Reflections on Big Science*, MIT Press, Cambridge.

WEISMAN, H. M. (1966) *Technical Report Writing*, Charles E. Merrill Books, Inc., Columbus.

WEISMAN, H. M. (1968) *Technical Correspondence: A Handbook and Professional Source for the Technical Professional*, John Wiley & Sons, Inc., New York.

WIGNER, E. P. (1950) The limits of science, *Proc. Am. Phil. Soc.* **94,** 422–427.

WILDHACK, W. A. and STERN, J. (1958) The Peek-a-boo system—optical coincidence subject cards in information searching. In Casey, R. S. *et al.*, *Punched Cards; their Applications to Science and Industry*, Reinhold Pubs., New York.

WILENSKY, H. L. (1967) *Organizational Intelligence: Knowledge and Policy in Government and Industry*, Basic Books, Inc., New York.

WOODFORD, F. P. (1967) Sounder thinking through clearer writing, *Science* **156,** 743–745.

WOOSTER, H. (1967) *As Long as You're Up, Get Me a Grant—The Preparation of Unsolicited Research Proposals*, Air Force Office of Scientific Research 65–0392.

Index